COMETS METEORS & ASTEROIDS–
HOW THEY AFFECT EARTH

Other TAB Books by the Author

No. 1505 *Understanding Einstein's Theory of Relativity: Man's New Perspective on the Cosmos*
No. 1525 *Black Holes, Quasars and Other Mysteries of the Universe*
No. 1605 *Basic Transistor Course—2nd Edition*
No. 1805 *Violent Weather: Hurricanes, Tornadoes and Storms*
No. 2000 *Encyclopedia of Electronics*

COMETS METEORS & ASTEROIDS—
HOW THEY AFFECT EARTH

STAN GIBILISCO

TAB BOOKS Inc.
Blue Ridge Summit, PA 17214

To Beverly

FIRST EDITION

FIRST PRINTING

Copyright © 1985 by TAB BOOKS Inc.

Printed in the United States of America

Reproduction or publication of the content in any manner, without express permission of the publisher, is prohibited. No liability is assumed with respect to the use of the information herein.

Library of Congress Cataloging in Publication Data

Gibilisco, Stan.
Comets, meteors, and asteroids—how they affect earth.

Includes index.
1. Comets—Popular works. 2. Meteors—Popular works.
3. Planets, Minor—Popular works. I. Title.
QB721.4.G53   1985   523.6   85-4621
ISBN 0-8306-1905-4 (pbk.)

Front Cover: Comet Ikeya-Seki just before sunrise. This comet, which appeared in 1965, had a long, spectacular tail and passed unusually close to the Sun. In general, the closer a comet's nucleus gets to the Sun, the brighter the tail becomes as the solar wind blows gas and dust millions of miles into space. Ikeya-Seki is one of a group of comets appropriately called Sungrazers. (U.S. Naval Observatory photograph.)

Contents

	Acknowledgments	vii
	Introduction	ix
1	**From the Beginning**	1

The Big Bang—Clouds of Gas—The First Generation of Stars—Creation of the Solar System—The Asteroids—Why Asteroids but Not a Planet?—Eccentric Orbits—Moons, Meteors, and Rings—The Comet Cloud—Major Impacts—The End of the Solar System—The End of the Universe—We Are Finite

2	**Fragments of Snow and Rock**	39

The Parts of a Comet—Early Theories about Comets—More Recent Ideas—Comet Spectroscopy—How Big Are Comet Nuclei?—The Solar Wind and Comet Tails—Different Comets Look Different—How Bright Are Comets?—Comets Change Appearances as They Move—Different Apparitions Will Look Different—Flare-Ups and Burnouts—Rotation of Comet Nuclei—The Effects of the Outer Planets—How Long Do Comets Last?

3	**Halley's Comet and Other Famous Comets**	73

Early Comet Apparitions—Characteristics of Memorable Comets—Naming of Comets—The Most Famous Comets—History of Halley's Comet—Orbit of Halley's Comet—What Will We See In 1985-86?—Visiting Comets—A Trip to a Comet's Head

4	**In Search of Comets**	107

Comet-Searching Devices—How to Comet Hunt—Astrophotography—Astrophotographic Comet Searching—Tracking—Why Search?

5 Stones from Space   **133**

Space Debris—Craters—Homemade Craters—Craters on the Earth—Meteor Showers—Ionized Trails—What Makes Up a Meteoroid—Meteoric Dust—Lagrangian Points—The Mystery of Tektites—The Asteroids—Finding and Tracking Asteroids—Hazards for Space Travelers

6 Catastrophism, Science, and Heresy   **175**

Scientific Method—A Modern "Heretic"—Catastrophism—Leaps and Bounds—Order and Disorder—Science and Reason—Democracy in Science—For the Sake of Argument—Do Comets Bring Life?

Epilogue   **199**

Bibliography   **201**

Index   **205**

Acknowledgments

I wish to thank the following institutions for their help in obtaining photographs for this book: the Smithsonian Astrophysical Observatory; Palomar Observatory, California Institute of Technology; Mount Wilson and Las Campanas Observatories, Carnegie Institute of Washington; the U.S. Naval Observatory; the National Aeronautics and Space Administration.

I have the deepest respect for those scientists and astronomers who devote their lives to further our understanding of the countless fragments of space debris that affect our lives.

Introduction

ON JUNE 30, 1908, A PIECE OF A COMET HIT the Earth. At least that is what seems to have happened. The event occurred in the Tunguska region of Siberia. Astronomers believe the fragment was large enough to have filled a football stadium.

Eyewitnesses described a large, bright fireball streaking across the sky from the southeast toward the northwest. The noise was heard for hundreds of miles from the point of impact. People as far away as 50 miles were blown off their feet by the blast wave. The results were similar to a hydrogen-bomb detonation, but that weapon had not yet been devised by humankind. It had to be a natural disaster.

Was it a volcano? No, because a cone-shaped mountain does not exist at the "ground-zero" point.

Was it a meteorite? Perhaps, but no large rocky or metallic fragment has been found at the site of impact.

Was it a piece of a comet? That is the most likely possibility. A fragment of snow and rock plunged toward our planet from interplanetary space and exploded with great violence in the atmosphere.

We are not sure exactly what a comet is or if all comets are alike. There are two basic theories:

☐ A comet could be a piece of ice, a mile or so across, with small rocky and metallic fragments embedded within it.

☐ Or a comet could be a swarm of meteoroids—pebbles—held together by weak mutual gravitation.

Credit for the first theory is generally given to the astronomer Francis Whipple, and the credit for the second theory is ascribed to the astronomer Raymond Lyttleton.

A third theory has been received with skepticism by scientists. Perhaps comets are ejected in violent eruptions by major planets—especially Jupiter. This theory is attributed to Immanuel Velikovsky. His theory is the most controversial, but perhaps the most interesting, of all.

We like to think of space as a more perfect vacuum than any evacuated chamber in any laboratory on Earth. In general, this is true. Except

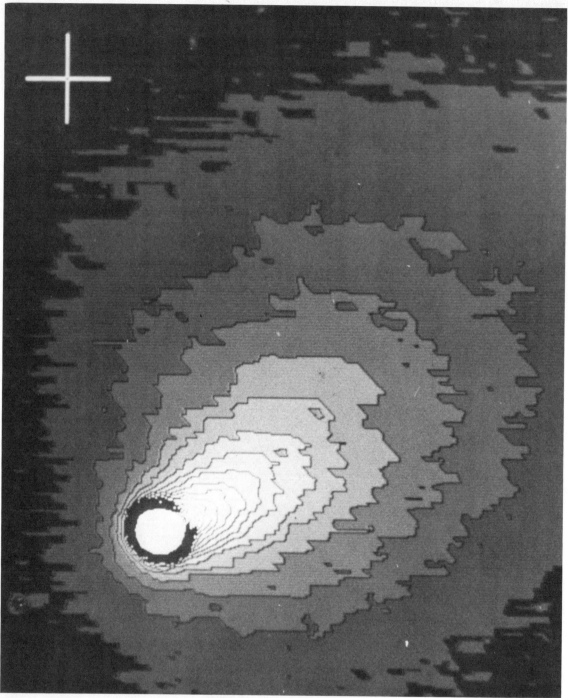

Fig. I-1. An image of the comet IRAS-Araki-Alcock with its nuclear structure enhanced by computer processing and the use of a charge-coupled-device camera. Various regions of luminosity are shown as discrete shades of gray. Note the brilliant jet of the tail near the coma. (Courtesy of Rudy Schild, Smithsonian Astrophysical Observatory.)

for the stars, planets, planetary moons, and a few octillion chunks of debris that hurtle through the void, space is indeed an almost perfect vacuum (at least compared to our planet's own atmosphere). Nevertheless, space is not quite empty between the orbits of the planets or among the wandering paths of the stars. Traces of all the natural elements, along with various compounds, can be found in space. Some of this material is in gaseous form and some is solid. The solid stuff makes up what we call comets, asteroids, and meteoroids. Interplanetary and interstellar comets, asteroids, and meteoroids are certainly more numerous than all the fish in all the seas of this world.

The Earth is constantly bombarded by particles no bigger than dust or sand grains. These tiny objects rain down by the millions each minute, but we do not notice them because they burn up in the upper air long before we can detect them. Perhaps every few tenths of a second, a pebble-sized piece of rock or metal enters the atmosphere and creates a brief flash, visible to those who might be watching on the dark side of our globe. Perhaps one in a thousand of these "falling stars" are really spectacular, leaving visible trails persisting for a minute or more. Sometimes a meteor explodes and shatters, and we see a fireball. Of these objects, a few actually reach the ground intact. People have seen meteorites and found them still hot. The largest meteorites blast craters in the Earth that last for thousands of years.

Where do these fragments of material come from? In order to know this, we must understand how the universe was formed and how it has evolved. This book begins by probing the history of the cosmos, all the way back to the beginning. Also described are the different theories concerning comets: their structure, formation, and behavior.

Some comets have attained special fame in times past. Halley's Comet, which returns every three-quarters of a century and has apparently done so for at least two millenia, is the most noteworthy example. There have been many other comets far more brilliant than Halley's. This book describes these comets and what they have taught us. Within these pages, you will be able to take an imaginary trip to the nucleus of a comet and ride with it as it swings around the Sun.

You will also find out how comets are discovered by amateur as well as professional astronomers, sometimes with just binoculars or small telescopes. Will you discover a comet and become famous?

Why is there no planet, but instead countless fragments of rock, between the orbits of Mars and Jupiter? What are the chances that one of these asteroids will go off course and careen toward the Earth? What would happen if a massive asteroid fell on our planet? We will find out these things, and discover how asteroids and meteoroids are formed.

In recent years, we have begun to learn that the universe might be less peaceful than we want to think. Things often happen by chance and are manifested as great disasters. One of the most interesting theories, put forth by Velikovsky in the middle of the twentieth century, holds that Venus and Mars once passed close to Earth and caused upheavals that are remembered to this day in the legends of many cultures. Whether or not Velikovsky's theory is true, we must accept that meteorites have fallen on the Earth, and certainly they have modified the geography, climate, and perhaps even the rotation of our planet. Our little world is probably not any safer from cosmic assaults than other parts of the universe.

Asteroids rain down on planets and their moons. Stars—and even whole galaxies—blow up. Planets, stars, and galaxies collide. It is thought that the universe was created in one gigantic explosion, more violent than anything imaginable. Catastrophe is the rule not the exception. This book examines how the theory of catastrophism is gaining acceptance.

Some day a comet or asteroid will collide with our planet and cause the whole world to shake, the skies to darken, and the oceans to overrun continental lowlands. Most likely, such an event will not occur within our lifetimes, and probably not within the lifetimes of our children and grandchildren. But the Earth is much older, and it will continue to exist for much, much longer than we.

A catastrophe *will* take place; it is only a question of when.

Chapter 1

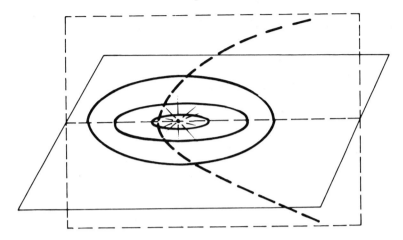

From the Beginning

PEOPLE HAVE ALWAYS ASKED TWO FUNDAmental questions about our universe:

- ☐ How is it put together?
- ☐ How did it come to be?

Modern astronomers and physicists have a good idea of the answers to these questions. Our knowledge of the universe has increased perhaps as much since 1900 as it did between that time and the dawn of humankind. We think the universe is like a gigantic, constantly expanding, four-dimensional sphere having a huge circumference—something on the order of 120 thousand million million million (120,000,000,000,000,000,000,000) miles. We figure that the universe is probably about 20 billion (20,000,000,000) Earth years old.

We flippantly throw these numbers around. We denote them in readable form by scrawling various hieroglyphic like symbols on a piece of paper. How well do we really understand the cosmos? Try counting to a million sometime. If you chatter off the numbers at one per second, it will take you 11 1/2 days. If you intend to count to a billion at the same rate, you will have to set aside almost 32 years. You will never count the number of years since the beginning of time and space. You cannot even begin to count the number of miles in the span of the known universe. The universe is too large and too old for us to comprehend. If we are to gain any understanding of our cosmos, we have to work with large numbers even if we cannot totally comprehend their magnitude.

Let us trace the development of the universe, galaxies, the Solar System, planets, asteroids, comets, and meteors from the beginning until now, according to the theories of contemporary scientists. We cannot say that the following story is absolute truth, but it is fairly well accepted.

THE BIG BANG

Originally, all of the matter of the universe was condensed into a space tinier than an atom . It might even have been infinitely small: a geometric point having no height, no width, and no thickness. The density of this primordial particle was beyond imagination. Perhaps it was infinite. Cosmologist George Gamow has given this bizarre substance a name—ylem—the stuff from which all things are made. Ylem was not matter as we know it. There were no atoms, no protons, neutrons, electrons, quarks, or other identifiable constituents. Nor was there radiant light or other energy in a form that we would call recognizable. Matter and energy were identical and inseparable. Time stood still.

Then, for some reason, the primordial particle blew up. Scientists do not know why it happened. Theologians give us as good explanation as anyone else: God caused it! Whatever the reason, the event did occur, and we can "hear" the repercussions with radio telescopes right up to this day. If we had been there to witness it, the explosion would have appeared fantastically brilliant. The first few sentences of the Bible paint a picture of it. The biblical description is hauntingly similar to what scientist tell us. Astronomers call the event the Big Bang. Theologians call it the moment of creation. Figure 1-1 depicts what it might have looked like during the first few instants in the life of our universe.

We can speak of seconds, minutes, hours, days, and years. This lets us get an idea of how the universe evolved. In the beginning, there was no Earth, no Sun, no clocks, and no people. We can't say with certainty, whether a second was like a year or a year like a second. All we know is that somehow time did pass. Time has continued to progress in such as manner as to make us believe that the Big Bang happened 20 billion of our years ago. That is about 250 million human lifetimes.

Immediately after the ylem exploded, it condensed into discrete particles. The particles at the outer edge of the expanding universe moved away from the center in all directions at almost the speed of light. They are still racing away at this rate. No matter can travel at the speed of light and still remain matter. Thus our universe has an edge, an outer limit, at which matter must become energy. This limit is 20 billion light years from our Solar System; that is the distance traveled by light since the birth of the cosmos.

Many different kinds of particles formed in the primordial universe, each particle having its unique characteristics. Physicists are still finding new kinds of particles today. Matter is exceedingly complex. There are so many types of particles that some philosophers suggest that the potpourri might be infinite.

Two kinds of particle were of special importance in the first few short hours of the primordial universe. We call these particles protons and electrons. As the universe expanded, it cooled, and electrons settled into orbits around the protons because the two types of particles had opposing electrical charges. This proton-electron union became the first element: hydrogen (Fig. 1-2). Hydrogen is still the most abundant element in the universe. All other elements have been, and are still being, made from it.

CLOUDS OF GAS

The universe grew rapidly larger, and it became less and less dense. Within a short time, the cosmos was a vast expanse of hydrogen gas clouds. The atoms bumped into each other with decreasing frequency. Ultimately the density of the universe was extremely low—almost a vacuum.

The force of the Big Bang was still felt: atoms were thrown about in great billows and wisps. This cosmic windstorm would have spent itself in darkness and cold had it not been for one crucial ingredient in the structure of time and space: gravitation. Something was trying to pull the atoms back together.

Gravitation, in general, was not strong enough to halt the expansion of the universe. The force of the Big Bang had been too great for that to be possible. Localized regions of the universe did acquire increased density. The atoms of hydrogen congregated in huge numbers into clouds of innumerable shapes. Some of the clouds were rotating (all of the atoms orbiting a common center). Other clouds simply hung in space. The centers of the clouds began to get dense as gravitation became stronger. The denser the congregation became the greater was the inward-working

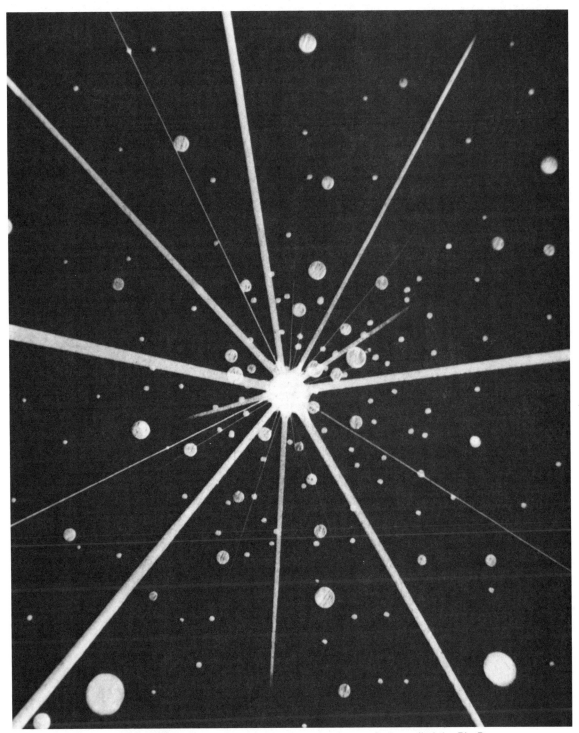
Fig. 1-1. Most astronomers believe that the universe began in a violent explosion called the Big Bang.

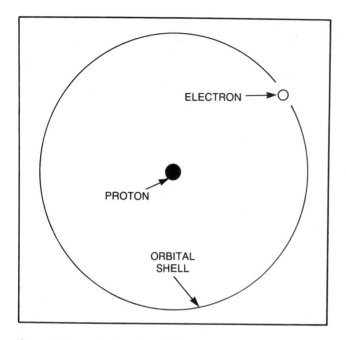

Fig. 1-2. The simplest atom is hydrogen, which is made of one electron and one proton.

force that was trying to pull them together. Atoms started colliding again. Heat was generated in the great clouds of hydrogen. The pressure mounted at the centers of the clouds.

The formation and evolution of galaxies is one of the least-known secrets of our universe. One thing works in our favor when we try to probe into the history of galaxies: the speed of light is the speed of time. The light that we see from any object, even a lamp 10 feet away, left it some time ago. We do not see distant stars and galaxies as they are now, but as they were years, centuries, or eons ago.

The nearest known star outside the Solar System appears to us as it was 4 1/3 years ago. The center of our Galaxy is about 40 thousand light years distant. The light we see from that region left before the beginning of recorded history. Using large telescopes and special photographing techniques, we can see objects that are several billion light years away. We can then peer into the past and get an image of a time when the universe was young, and the Earth and Sun did not yet exist.

There is a limit to the universe we can see with even the largest optical telescopes, but there is another method of observation available to us: the radio telescope. Pioneered by various astronomers in the middle of the twentieth century, radio telescopes allow us to gaze even further back than we can see with visual apparatus. We can use the radio telescope to observe the primordial fireball—which appears as a faintly glowing sphere around us—with its "surface" some 20 billion light years and receding from us at almost the speed of light.

There are millions upon millions of galaxies within the range of optical and radio telescopes. An average galaxy, such as our Milky Way, contains about 200 billion stars. Some elliptical galaxies have as many as a trillion individual stars; small galaxies might have only a few million. A casual observer might look at photographs of various galaxies and conclude that they all are pretty much alike, with bunches of stars of greater or lesser size. Astronomers have ways of detecting subtle differences.

Every object in the universe emits some radio energy. The galaxies are no exception. When radio telescopes are used to "look" at galaxies at radio wavelengths, differences are noticed. Some galaxies are "radio dim" while others are "radio bright," Nevertheless, the "radio brightness" of a galaxy is not highly correlated with the visual brightness.

Some of the largest and optically brightest galaxies put out very little energy as radio waves, while some of the optically dimmest galaxies generate enormous energy in the radio spectrum. The latter type of galaxy is appropriately called a radio galaxy. Most of the radio galaxies are very far away—billions of light years from us. Even more remote are the mysterious quasars. They generate as much energy as a whole galaxy, but quasars are just a tiny fraction of the size of the Milky Way.

Because the quasars are all very far away, we see them all as they were long ago. This leads many astronomers to believe that quasars are protogalaxies (galaxies in the process of formation).

The following is one possible theory for the formation of the Milky Way Galaxy billions of years ago. A cloud of hydrogen gas, swirling around a gravitational center, began to heat up. The warming took place first at the relatively dense center. Within the great cosmic whirlwind of hydrogen gas, smaller eddies developed, just as tornadoes form within larger revolving systems in the atmosphere of our planet. At the centers of these cosmic tornadoes, the hydrogen became so dense and hot that nuclear fusion began. For the first time since the Big Bang, there was light! Similar events happened all over the universe.

At first, only the center of the hydrogen cloud had stars. Stars developed later in the outer regions of the Milky Way. Because of this, we find the oldest stars in our Galaxy near the center, and younger stars near the periphery. Today, our galaxy glows with starlight throughout its spiral of matter. Figures 1-3A and 1-3B show what our Galaxy would look like as viewed from a million light years above its plane (A) and from a half million light years edgewise (B).

THE FIRST GENERATION OF STARS

The hydrogen in the young Milky Way did not all aggregate at the center. There is, however, some evidence to suggest that most of it has condensed there. The center of our Galaxy was where the first nuclear fusion reactions took place. These reactions started the element-building processes that eventually led to the existence in nature of 91 other types of atoms in addition to hydrogen.

When hydrogen is compressed sufficiently, it gets so hot that strange things happen to its atoms. First, the electrons are stripped from their orbits around individual protons (Fig. 1-4). Instead of staying with one mate, any given electron moves freely around among the protons. Finally the alchemy begins; hydrogen is changed into helium. Four protons merge together to create a new kind of atomic nucleus. In the process, two of the protons lose their electric charge, creating a previously nonexistent particle known as a neutron. A little bit of the matter is converted into energy: radio waves, heat, light, ultraviolet, X rays, gamma rays, and an assortment of high-speed particles. The hydrogen fusion process, which is the first step toward the formation of heavier elements, is illustrated in Fig. 1-5.

Ancient astronomers believed that our Sun shone because it was a huge bonfire. Even as recently as the nineteenth century, physicists did not really know how stars worked. Modern scientists believe that it is the energy of atomic fusion that makes a star shine so brightly. *Stars operate according to the same principle by which hydrogen bombs explode.

Nuclear fusion can occur not only among hydrogen atoms, but also among helium atoms. In this way, heavier and more diverse elements—oxygen, iron, carbon, nitrogen, and all the rest were formed inside the first stars—until ultimately there were some atoms with as many as 92 protons.

Gradually, stars formed farther and farther away from the center of the Galaxy. We still see some of those stars today. Others we do not see because they became unstable and blew up. In some galaxies, it is possible that many, or most, of the stars in the center blew up all at once. Such dramatic stellar deaths still occur. An exploding star becomes millions of times brighter than before.

---

*A detailed discussion of stars and galaxies is given in *Black Holes, Quasars and Other Mysteries of The Universe* (TAB Book No. 1525).

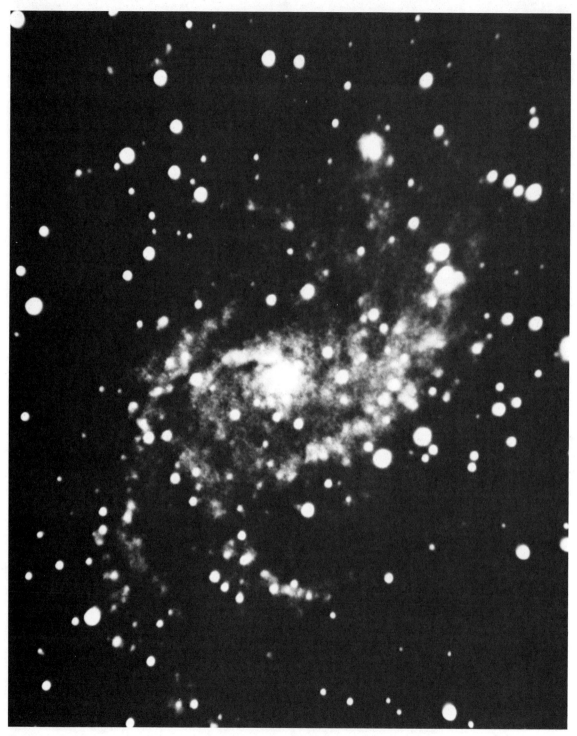
Fig. 1-3A. Our Galaxy, seen from far away, would look similar to this pinwheel of stars.

Fig. 1-3B. Seen edge on, the Milky Way would appear like this spiral (U.S. Naval Observatory photographs).

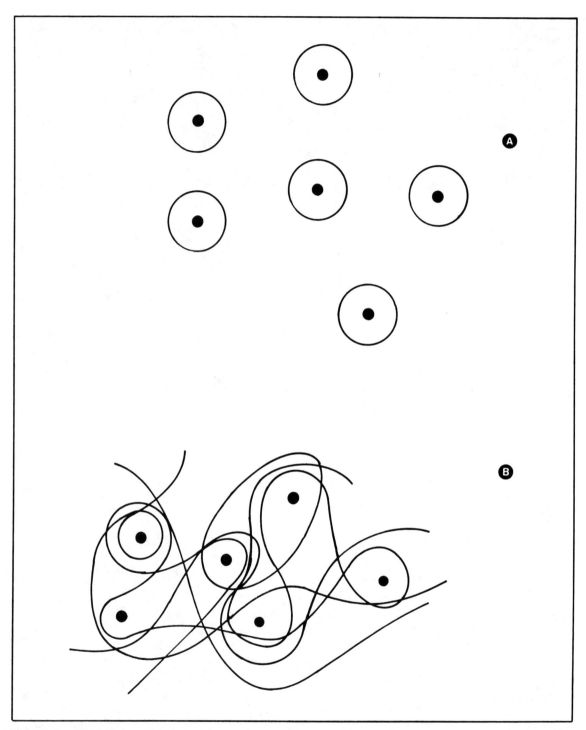

Fig. 1-4. At a relatively low temperature, electrons tend to stay with single nuclei (A). But when things get hot, electrons wander (B).

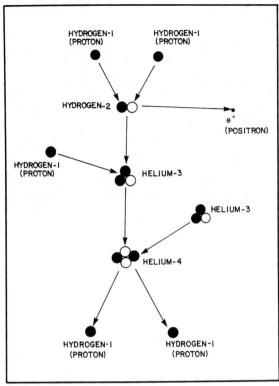

Fig. 1-5. The hydrogen fusion reaction is what causes stars to give off energy.

Sometimes we can see a steller explosion, a supernova, from Earth.

Exploding stars leave entrails—clouds of gas, dust, and chunks of materials of varying size—in their vicinity. When we see these clouds through a telescope, some appear to glow faintly while others appear dark, obscuring the stars behind them. Most of the clouds form near the plane of our galaxy and other spiral galaxies. Dust clouds are clearly visible in other galaxies when they present themselves to us edgewise (as shown in Fig. 1-3B). Our Galaxy is thick with these clouds. This dims the light that arrives from the center of the Milky Way. Many of these clouds can be seen through a telescope in the direction of the constellation Sagittarius. Figure 1-6 shows one such view. If there were no dark clouds in our Galaxy, the center would appear almost as bright as the Sun.

Some astronomers believe that it is in these clouds that most of the comets, meteors, and asteroids exist. It is from these clouds that second-generation stars, such as our Sun, are born. Figure 1-7 shows a region in which new stars are condensing from such a cloud. The Earth and the other planets in our Solar System formed from interstellar gas, dust, comets, meteors, and asteroids. All of the metals and rocks in the Earth ultimately came from this material. So did the water in our oceans, and all of the plants and animals that populate this planet. Our bodies are made of the same atoms that once were the constituents of interstellar debris!

CREATION OF THE SOLAR SYSTEM

Our Sun is a star of the second or perhaps even third generation. Our parent star formed from the remnants of older stars that became supernovae. The Sun was not the only thing produced by that cloud; nine daughter planets (that we know of) and innumerable moons, asteroids, meteors, and comets orbit the central star.

Why should the Sun have satellites? Why isn't it just standing alone in space? There are two major theories about how the planets and other nonstellar objects came to be. The two models are called:

☐ The tidal or maverick-star theory.
☐ The rotating-cloud or ring theory.

The Tidal Theory

According to the tidal theory, the Sun originally had no planets or other satellites. Our parent star actually did form alone, and the other objects in the Solar System came later.

The Milky Way galaxy is 100 thousand light years across, but it contains 200 billion stars, all revolving around the nucleus like an enormous swarm of honey bees. Not all of the stars follow perfectly circular paths in the plane of the Galaxy. Some stars have eccentric orbits and some, like the Sun, bob up and down, passing through the plane of the Galaxy at regular intervals. Close encounters

Fig. 1-6. The center of the Milky Way Galaxy lies in the direction of the constellation Sagittarius (U.S. Naval Observatory photograph).

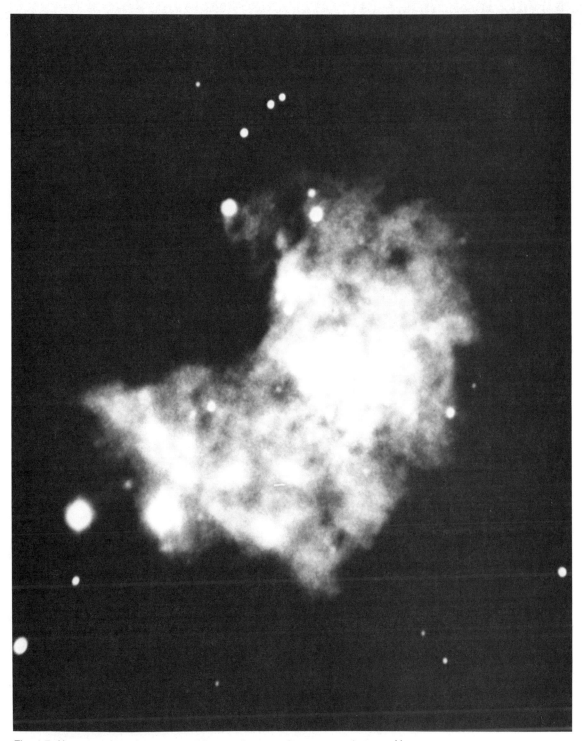

Fig. 1-7. New stars form from gas and dust (U.S. Naval Observatory photograph).

between different stars can, and most certainly do, happen.

Suppose that another star came extremely close to the Sun—say within a few hundred thousand miles. The gravitational tug-of-war would be tremendous. Both stars would be visibly distorted, stretched into football shapes. The two stars might pull at each other with such force that an exchange of stellar material would result (Fig. 1-8). The Sun might pull matter out of, and give up some of its own matter to, the other star. As the stars swung around and parted ways, most of the starstuff would be pulled into the two members, but some would scatter into orbits around them. Perhaps the planets, asteroids, meteors, and comets formed from this stellar matter as it cooled and condensed.

Contemporary astronomers doubt that this is the way the Solar System was born, and they have powerful arguments to support their contentions. If the Earth and other planets, in particular, were generated in this manner, we would expect that their paths around the Sun would describe highly irregular orbits. These orbits would almost certainly not lie in the same plane, and would not necessarily progress in the same direction around the Sun. Yet, all of the planets orbit in nearly perfect circles in an almost common plane. All of the planets revolve in the same direction as the Sun spins on its axis. If the tidal theory is correct, then this uniformity is a great coincidence.

The Rotating-Cloud Theory

The Sun turns on its axis and makes one complete rotation about every month. It thus makes sense

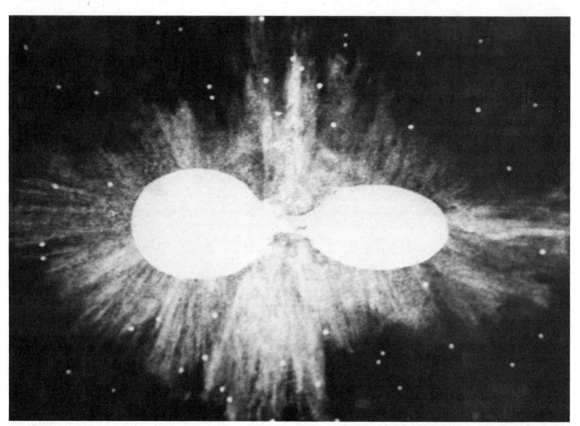

Fig. 1-8. If another star passed very close to the Sun, matter might be pulled from both stars. Some scientists think that this is how the Solar System was created: the matter condensed to form the planets, moons, asteroids, meteoroids, and comets.

Fig. 1-9. The most commonly accepted theory of the formation of the Solar System holds that rings of gas and dust congealed to form the planets.

to think that the original debris, from which the Sun was made, was not just hanging in space but was spinning. Every atom, dust grain, comet, meteor, and asteroid was orbiting around a gravitational center.

Astronomers have shown that a cloud of debris, being pulled inward by gravitation, would almost certainly rotate around a common center. The debris would align itself approximately in a plane. Perhaps the most interesting thing of all is that the debris would tend to get concentrated into discrete, nearly circular rings (Fig. 1-9). This is a natural result of the spinning momentum of the central Sun. If this did not happen, the Sun's rate of spin would be much faster than it actually is—so much faster that the Sun might literally fly apart.

As millions of years passed, according to the rotating-cloud theory, gravitation pulled the debris in the rings together, forming massive, solid bodies: planets and moons.

The rotating-cloud theory explains why the planets orbit the Sun in such a uniform manner. Moreover, this theory provides us with the hope that there are other Solar Systems in our Galaxy. Because the formation of rings is necessary according to the laws of physics, it follows that other stars, not just the Sun, would have ring systems from which planets could develop.

Recently, astronomers found evidence of such a ring system around the star Vega in the constellation Lyra. Vega is a relatively nearby white star about 50 times as bright as the Sun. The discovery was made quite by accident, using infrared detecting apparatus.

If Vega really is surrounded by rings—planets in the process of forming—it means that our Solar System is not a freak cosmic accident. That tells us there are probably millions or even billions of other solar systems in our galaxy. Some of these solar systems might have Earthlike planets, perhaps with intelligent life.

THE ASTEROIDS

In the eighteenth century, the German astronomer Johann Bode found that there exists a strange relationship among the distances of the planets from the Sun. (Bode's law had been discovered previously by German astronomer J.D. Tietz.) Bode's sequence is generated starting with 0, then 3, and doubling the value each time: 6, 12, 24, and so on. Each number is then increased by four, and the results divided by 10. This gives the distance from the Sun to each planet in astronomical units. (An astronomical unit is 93 million miles, the radius of the Earth's orbit around the Sun.)

Table 1-1 compares the actual mean orbital radii of the planets with the numbers of Bode's sequence. There is one flaw in an otherwise almost perfect correlation from the orbits of Mercury through Uranus. No planet exists for the Bode orbit just past Mars.

Of course, we might think that Bode's sequence is nothing but a coincidence. It seems odd to suggest that solar systems would arrange their planets by performing arithmetic operations on products of 2 and 3! Nevertheless, Bode believed that there was some reason for the correlation, and some other astronomers agreed with Bode.

A search was begun for a planet between Mars and Jupiter. An Italian, Giuseppi Piazzi, discovered

Table 1-1.

Ordinal	Bode Number	Planet	Distance from Sun, Astronomical Units
0	0.4	Mercury	0.387
1	0.7	Venus	0.723
2	1.0	Earth	1.00
3	1.6	Mars	1.52
4	2.8	None	-
5	5.2	Jupiter	5.20
6	10.0	Saturn	9.54
7	19.6	Uranus	19.2

the missing planet at the beginning of the nineteenth century. But it was tiny—less than 500 miles across—when compared with any of the other known planets. Piazzi named the object Ceres after the Roman goddess of agriculture.

Ceres became the first known asteroid. Asteroid means "starlike," and the term is well chosen. All asteroids, like stars, appear as points of light—even in powerful telescopes. Asteroids are sometimes called planetoids, meaning "planetlike." This is also a good name for the objects, because, like planets, they shine by reflected sunlight—not by their own light.

Bode was vindicated by the discovery of Ceres, but it was somewhat bothersome that the planet was so small. Astronomers were even more puzzled when more asteroids were found having orbits similar to that of Ceres. Since Piazzi found the first asteroid, thousands more have been discovered. Most of them have orbits roughly corresponding to that predicted by Bode. The asteroids are arranged in a relatively uniform ring around the Sun.

WHY ASTEROIDS BUT NOT A PLANET?

Why did no planet form in the Asteroid Belt? Why was this ring of debris unable to pull itself together? There are several possible reasons.

All of the asteroids put together would not make a very large planet, according to modern estimates of the amount of material in orbit between Mars and Jupiter. Perhaps the gravitation among the fragments was insufficient to allow a planet to form. Star-system rings might not always pull themselves together to make planets. Apparently, a ring must contain a certain minimum amount of matter in order to form a planet. If there is not enough mass, the ring remains in pieces.

Some scientists have suggested that the asteroids are the remnants of a planet that once existed. If this is true, then the planet was smashed to bits. Why would that happen? Perhaps the gravitational field of Jupiter exerted tidal forces that the planet could not endure. Perhaps a large comet or an asteroid collided with the planet.

A less likely, but haunting, possibility is that the planet was populated by intelligent beings who fought a war with great bombs, destroying their planet along with themselves. We humans can get some comfort from the fact that the planet, if it did once exist, was probably too small to hold an atmosphere and too cold to support life as we know it. Therefore, this theory is probably not correct!

ECCENTRIC ORBITS

The orbits of the planets, as well as most of the asteroids, are almost (but not quite) perfect circles. Some celestial objects move through space along routes that do not even remotely resemble circles. Some orbits aren't even closed; the objects do not stay in the Solar System.

All of the planets follow orbits that are elliptical (not circular). Even our own planet's distance from the Sun varies over a range of about 2 percent, although this is not enough for us to notice without precision equipment. With the exception of Pluto, the same is true of the other planets. Pluto's orbit differs considerably from a perfect circle.

The shape of the Asteroid Belt is, in general, almost circular, but a few individual asteroids wander far outside the main path. Some of them cross inside the orbit of Mars or outside the orbit of Jupiter. Eros comes within 14 million miles of our planet. A few asteroids occasionally get closer to the Sun than the Earth. Adonis and Apollo are two examples of such asteroids. These objects might actually collide with our planet eventually, although the chance of it happening soon is extremely small.

A chunk of cosmic flotsam called Icarus ventures closer to the Sun than the planet Mercury. This asteroid gets its name from the Greek legend of Icarus and Daedalus. Two men of ancient times built wings by waxing bird feathers together. Their intent was to fly through the air. They were successful, and soared like birds to great heights! Daedalus knew the danger of flying too high; the heat of the Sun might melt the wax that held the feathers together.

We know today that this must be a legend and not a true story. The temperature of the air drops with increasing height above the ground.

Icarus, in his foolhardiness, flew too close to the Sun with his wings of waxed feathers. The heat from the Sun melted the wax. Icarus spiraled down into the sea to his death.

The asteroid named Icarus does indeed roast in the heat of the Sun, dipping to within 20 million miles of our parent star. That is scarcely more than half the orbital radius of the planet Mercury. When Icarus is at perihelion, it receives more than 20 times as much sunlight as the surface of the Earth.

Why do some of the asteroids have such errant orbits? Astronomers believe that asteroids occasionally smash into each other. Such a collision would knock one or both of the objects from an almost circular orbit into an eccentric orbit. Perhaps some asteroids have been sent plunging into the fires of the Sun as a result of such collisions. Others, like Adonis and Apollo, are thrust into orbits that might someday threaten life on Earth.

Some celestial objects move with such eccentricity that we might be tempted not to think of their paths as orbits at all. Such fragments of debris are not permanently under the gravitational influence of the Sun. They come from interstellar space, pass near the Sun, and depart never to return. Their routes through space are hyperbolic.

An astronomer named Johannes Kepler

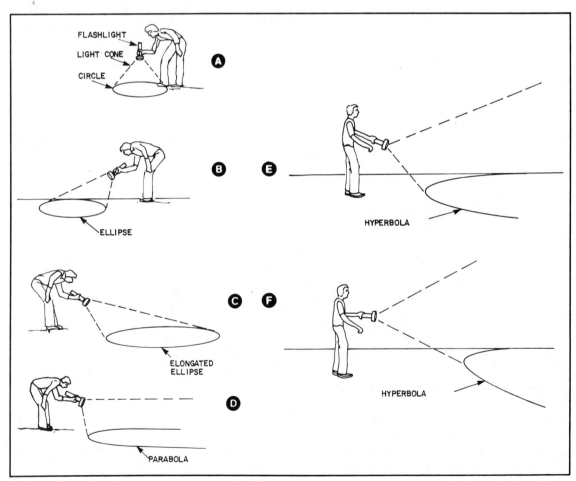

Fig. 1-10. Conic sections can be "drawn" with a flashlight: a circle (A); ellipses (B and C); a parabola (D) and hyperbolas (E and F).

showed that all celestial must be conic sections: either a circle, ellipse, parabola, or hyperbola. The curves obtain their names from the fact that they are cross sections of cones. You can create your own conic sections by shining a flashlight onto a flat surface at various angles, as shown in Fig. 1-10.

As you stand on a vacant parking lot or other flat, open place on a dark night and shine your flashlight upward, the outline of the large, dim part of the light beam (not the bright central ray) is cast into space in the shape of a cone. You can consider this cone to be infinitely long (although it is actually not because light travels at a finite speed.).

The Circle

Suppose you direct the flashlight straight downward. Then the ground cuts off the cone of light in the shape of a circle (A of Fig. 1-10). This represents a "perfect" orbit, taken only by satellites that are never influenced by the gravitational fields of other objects.

A celestial object orbiting in a perfect circle would maintain absolutely constant distance from the Sun at all times. The orbital radius would not vary by a mile, a foot, or even an inch! In theory this can happen; in real space it does not. Such a state of perfection is restricted to the realm of mathematics. In real life, the probability of its happening is zero. And even if, by some coincidence, a circular orbit did exist, something would come along sooner or later and collide with the object and throw off the orbit. For these reasons, no planet or asteroid follows a perfectly circular orbit around our parent star.

The Ellipse

If you tilt the flashlight a little, the outline of the dim part of the beam lands on the ground in the shape of an ellipse (B of Fig. 1-10). As you tilt the flashlight further, the ellipse becomes more elongated (C of Fig. 1-10) as the far edge moves further and further away from you. You can create an ellipse, having any amount of elongation you want, by tilting the flashlight more or less. You can also change the nature of the ellipse by moving the flashlight higher or lower above the surface.

An ellipse is not as mathematically unique as a circle. Celestial objects commonly describe elliptical orbits. A satellite in this kind of orbit moves alternately closer to and farther from the Sun. The object travels fastest when it is is nearest the Sun and slowest when it is farthest away.

Elliptical orbits are defined in terms of eccentricity, an expression of the extent to which the orbit differs from a perfect circle. A perfect circle has eccentricity zero; progressively more elongated ellipses have larger and larger eccentricity values.

The Parabola

If you tilt the flashlight until the far edge of the ellipse just leaves the ground, the dim outline lands in the shape of a parabola (D of Fig. 1-10). A parabola is an open curve.

Like the circle, the parabola is a mathematically special curve, and no celestical object ever exactly follows an orbit of this shape. A satellite in a parabolic orbit has just enough speed to achieve escape velocity from the Sun: no more and no less. This can occur in theory but not in reality.

Although some highly eccentric elliptical orbits seem parabolic near the Sun, small differences invariably exist.

The Hyperbola

When the flashlight is tilted a little further still, the dim outline describes a hyperbola (E and F of Fig. 1-10). Like the ellipse, the hyperbola can be generated in an immense number of different shapes and sizes. It is not as specialized as the circle or parabola. When an interstellar wanderer enters the Solar System, swings around the Sun and leaves forever, its orbital path is a hyperbola.

Elliptical and hyperbolic orbits can change in shape or size because of the gravitational effects of the planets. Jupiter, with its powerful pull, can capture an object from outside the Solar System and change the object's orbit from a hyperbola to an ellipse (Fig. 1-11). Saturn, Uranus, and Neptune occasionally are responsible for similar events. If an asteroid passes especially close to one of the

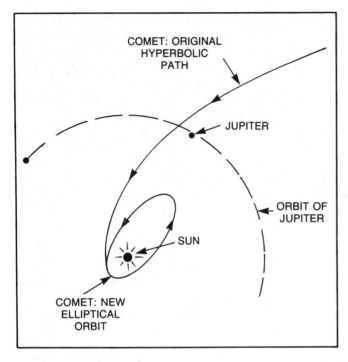

Fig. 1-11. An object entering the Solar System, along a hyperbolic path, might be deflected by Jupiter and thrown into an elliptical orbit—thus being captured permanently.

giant planets, the wanderer sometimes becomes a satellite of that planet: a moon. Jupiter and Saturn have numerous visible moons—and probably hundreds or even thousands of tiny moons—swept up from debris that originated both inside and outside of the Solar System.

MOONS, METEORS, AND RINGS

As the planets formed in the Solar System, not all of the matter became concentrated into single objects. Some planets formed in pairs or groups. Our Earth-Moon system is considered by some astronomers to be a double planet. Pluto is another example of a planet with a single large moon. Some planets, such as Jupiter, congealed along with several large moons.

The Solar System is a vortex within the galaxy, rotating fast near the center and slowly in its outer reaches. The Sun spins around about once a month. Mercury orbits the Sun in 88 days. Pluto, the most distant known planet, requires almost 250 years to complete one revolution.

According to the rotating-cloud theory, the planets formed as the Sun was forced to shed some of its spin, a phenomenon that physicists call "the conservation of angular momentum." Planet-moon systems are smaller vortices; they have periods ranging from a few hours to about a month. Moons probably formed around the planets for the same reason planets developed around the Sun: a transfer of momentum among individual particles of matter.

The planets in the Solar System orbit in a plane that is greatly tilted with respect to the plane of the Milky Way spiral. The ecliptic, or the plane of the Earth's orbit around the Sun, is tilted almost 90 degrees to the galactic plane. Most of the planetary moons also orbit in this tilted plane.

In most instances, the moons probably formed at the same time as, and in conjunction with, their parent planets. Our double planet is one example. Our Moon is quite large; it is almost half the diameter of the Earth. It is unlikely that the Moon was once a separate planet and was then captured by the Earth's gravity, but some moons might be former asteroids. Phobos (Fig. 1-12A) and Deimos (Fig. 1-12B), the tiny moons of Mars (Fig. 1-12), might once have been part of the asteroid belt, eventually falling into the gravitational grip of

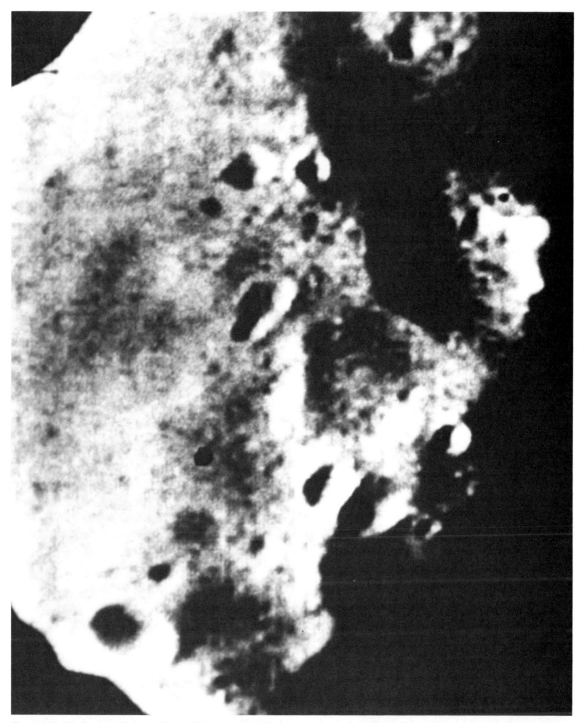

Fig. 1-12A. Phobos and Deimos (Fig. 1-12 B), moons of Mars, might be former asteroids that were captured by the gravitation of the planet (courtesy of NASA).

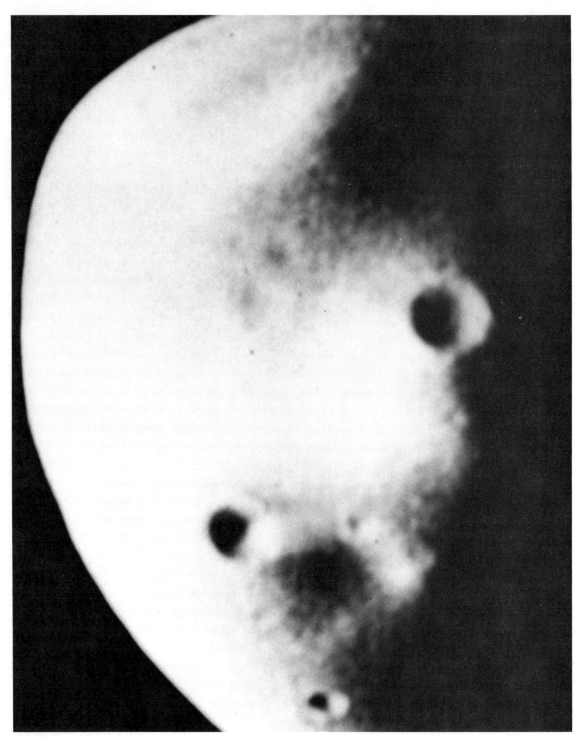

Fig. 1-12B. Deimos (courtesy of NASA).

Mars. Hyperion, a moon of Saturn (Fig. 1-13), is another example of a possible "asteroid" moon.

"Natural moons," or satellites that formed along with their parent planets, appear spherical like planets. "Asteroid moons" look like asteroids; they are smaller and more irregular in shape. "Natural moons" generally orbit their parent planets in the same direction the planet rotates: west to east. "Asteroid moons" could orbit in the opposite direction. Jupiter and Saturn probably have many undiscovered "asteroid moons." Even our planet might have numerous tiny satellites of this type.

"Natural moons" usually orbit their parent planets close to the plane of the planet's orbit around the Sun (A of Fig. 1-14). Our own moon deviates only slightly from the ecliptic. The moons of Uranus, however, are an exception. They orbit the planet in the plane of its equator, which is tilted almost 90 degrees from its orbit around the Sun (as shown in B of Fig. 1-14). "Asteroid moons" might orbit in any orientation—even over the poles of the planet—depending on the circumstances of their gravitational capture.

Our Moon provides us with graphic evidence that the Solar System is peppered with asteroids of all sizes. The Moon's surface is heavily cratered (Figs. 1-15A and 1-15B). The Moon has no wind, or rain to cause erosion, and the craters last for many millions of years. On Earth, craters are covered up by vegetation or worn down by erosion. Because of this, we do not find very many craters on our planet. The Earth is larger than the Moon, and the Earth has a stronger gravitational field. Our planet has been—and will continue to be—hit by asteroids more often than the Moon. Some of the these impacts have had a significant effect on our planet and on the creatures who dwell here.

We can derive some comfort from the thought that there are probably fewer asteroids near the Earth's orbit today than there were when the Solar System was young. The Earth and Moon gradually swept them up in repeated passages around the Sun. The same thing has happened near the orbits of the other planets. The chance of a major asteroid impact is smaller now than it was a billion years ago, but the probability is not quite zero!

There are almost certainly many asteroids still orbiting the Sun in between the planets. Interplanetary travelers will have to look out for them. Once in a while, two asteroids collide and send a chunk of rock flying toward one of the major planets or moons. The asteroid is then either captured, becoming a new moon of the planet, or strikes the planet or one of its moons. A crater would result from a collision and perhaps there would be a major geologic upheaveal with repercussions lasting for centuries.

Small asteroids, more commonly known as meteoroids, are abundant in interplanetary space. A meteoroid is known as a meteor if it enters Earth's atmosphere. Most meteors burn up in a blaze of heat and light before they can reach the ground. On clear nights, falling meteors can occasionally be seen, glowing brightly and leaving a brief trail. Such occurrences have been called shooting stars or falling stars.

Meteors tend to fall in batches known as meteor showers. During an intense meteor shower, you can see several shooting stars each minute. Rarely do they appear at the rate of one or more

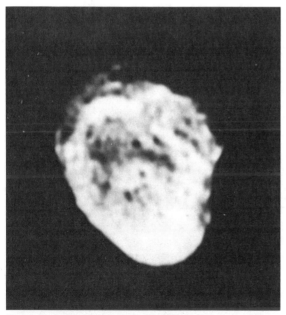

Fig. 1-13. Hyperion, a small moon of Saturn, is perhaps a former asteroid (courtesy of NASA).

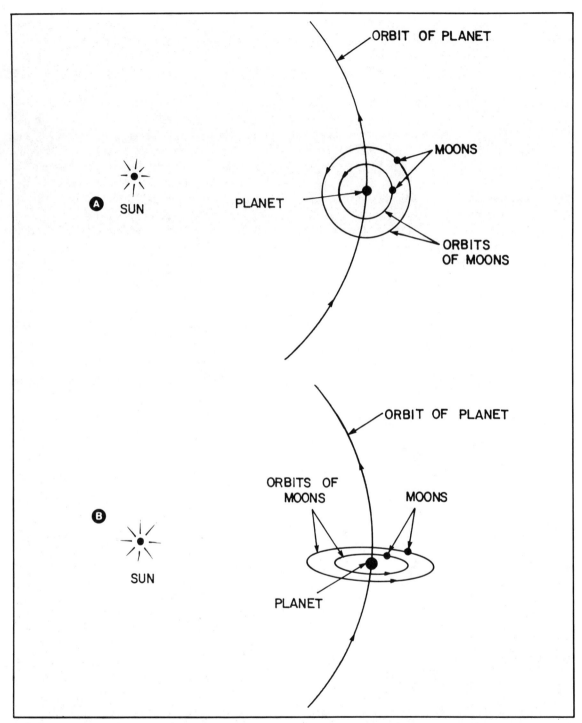

Fig. 1-14. In most cases, the moons of a planet orbit in almost the same plane as the planet itself (A). But in some instances this is not true (B).

Fig. 1-15A. The pockmarked surface of the full Moon gives evidence that space is filled with debris.

Fig. 1-15B. A closeup view of the Moon (courtesy of NASA).

per second. Meteor showers occur regularly throughout the year; the most common showers are listed in Chapter 5.

Some meteors are large enough to survive the challenge of the atmosphere and they strike the surface. Then they are called meteorites. Some people have actually witnessed the fall of a meteorite. There are even some cases on record in which houses have been directly struck! There's no need for any of us to get overly concerned about meteorites falling on our property. The probability of a direct hit is exceedingly minute.

Moons, either "natural" or "asteroid," are not the only satellites that planets can have. We know this immediately when we use even a small telescope to observe Saturn. Saturn is surrounded by millions upon millions of tiny moons in concentric, almost perfectly circular, orbits near the plane of the equator. These are the famous rings (Fig. 1-16) of Saturn. Space probes have recently found that Jupiter and Uranus also have rings, but they are less prominent and do not show up through Earth-based telescopes.

Planetary rings are made up of icy particles. Such materials are much different from asteroids. The reason for the existence of rings is still largely a mystery. Some astronomers think that the rings of Saturn result from matter pulled from one of the moons, such as Titan (which has an atmosphere), into orbit around the parent planet. The rings form as the material slowly spirals inward, ultimately falling onto Saturn itself. Other scientists theorize that the rings were once a small moon that was torn apart, by the powerful gravitation of Saturn, and gradually scattered uniformly around the planet. Another theory suggests that the rings would have become a moon, but were prevented from aggregating by Saturn's gravity or because there was not enough material. The rings would then be a sort of miniasteroid belt. Still another possibility is that the rings are the remnants of a comet that ventured too close to the planet and was ripped asunder by tidal forces.

THE COMET CLOUD

Comets are one of the greatest mysteries in astronomy. They have awed and terrified men since the beginning. Even today, many people believe that comets have some kind of supernatural characteristics. Twentieth-century people, almost as much as the ancients, fear or dread the apparition of a comet. Nevertheless, as one astronomer has said, comets in our Galaxy are more numerous than fishes in the sea! And they are no more or less threatening.

According to Dr. Fred Whipple, comets revolve around the Sun in a vast spherical halo—far beyond the orbit of Pluto—where the Sun is hardly brighter than a distant star. In this perpetually frigid environment, chunks of rock and ice meander slowly, just barely under the influence of the gravitation of our parent star (Fig. 1-17).

According to the Whipple model, a star occasionally passes fairly close to the Solar System, upsetting the normally regular orbits of the objects in the distant comet cloud. Some of the comets are sent into orbits so irregular that they pass extremely close to the Sun, becoming visible to us through binoculars or even to the unaided eye. The disturbance might be caused by different stars at irregular intervals or by a single star whose motion carries it periodically toward and away from the sun.

It has been suggested by some astronomers that the Sun has a smaller, companion star that upsets the comet cloud every few million years. The result is that a barrage of comets is hurtled at the Sun. Some of the comets strike our planet and cause changes in the environment. Because climatic change (whatever its causes) has perhaps been responsible for such cataclysms as the extermination of the dinosaurs, this hypothetical sister star is sometimes called the death star. While that term sounds a little sensational, at least it grabs attention.

Another theory of cometary origin suggests that comets are interstellar wanderers that drift through space, waiting to be captured by the gravitation of a Solar System. More or fewer comets would then be observed as the Sun plunges back and forth through the plane of the Galaxy. If this theory is correct, we would observe (over intervals of millions of years) fluctuations in the number of

Fig. 1-16. The rings of Saturn (courtesy of NASA).

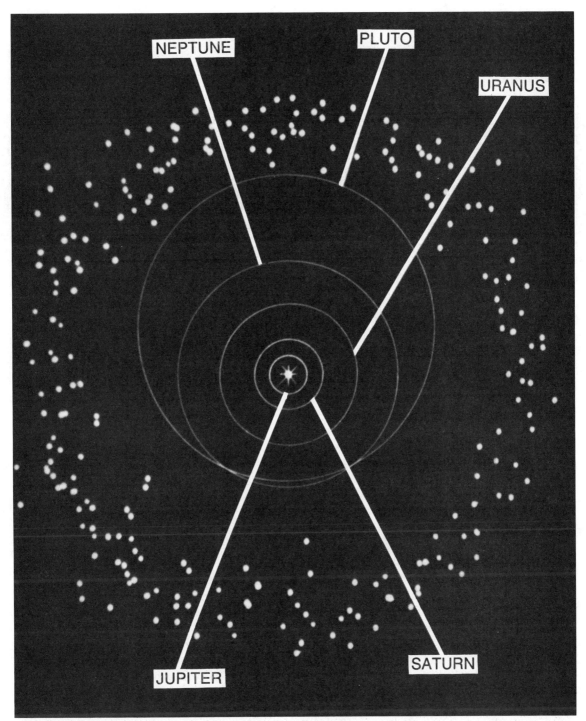

Fig. 1-17. Some astronomers believe that comets swarm around the Sun in a vast halo beyond Pluto. Passing stars occasionally hurl some of the comets toward the Sun.

comets (just as would result from the action of a death star).

Some solar researchers think that the Sun has thrown material into space while going through an unstable phase called a T.Tauri period. According to this theory, material could have been flung past the orbit of Pluto and condensed into comets having irregular orbits. It is true that the Sun is much less stable than we thought a few decades ago, and there is some good evidence that our parent star changes significantly in brightness and warmth. This is a disturbing realization. We rely on a constant, well-tempered Sun for our very existence.

A fairly new theory proposes that the comets were formed when a planet, orbiting between Mars and Jupiter, blew up just a few million years ago. According to this idea, asteroids, meteoroids, and comets all came from this planet. If the asteroids are the remnants of a planet—some astronomers have suggested that they are—then perhaps the comets were formed along with the asteroids.

Probably the most controversial theory of comet origin is that they come from large planets. This idea, first put forth by Immanuel Velikovsky in the middle of the twentieth century, proposes that comets have been ejected by the large planets—especially Jupiter. Velikovsky believed that a massive comet, much larger than any in recent times, was thrown into space by Jupiter in a violent eruption. It has been suggested that the Great Red Spot, a disturbance on Jupiter that is startlingly similar in diameter to the planet Venus, might be the entrails of that eruption (Figs. 1-18A and 1-18B). The object passed near the Earth and Mars, causing geologic and climatic changes on our planet. Finally the "comet" settled into a nearly circular orbit about halfway between the orbits of Mercury and the Earth. Now we call it Venus!

What are the comets made of? Ancient astronomers and philosphers had a variety of theories. After the death of Julius Caesar, emperor of the Roman Empire who was murdered in 44 B.C. a spectacular comet appeared. This comet was considered to be the spirit of Caesar being transported into the heavens. Aristotle thought that comets were caused by matter ejected from the Earth into the atmosphere. Today we have what we think are more accurate concepts.

There are two major contemporary theories concerning the structure of comets.

☐ Fred Whipple believes that comets are large balls of icy matter, peppered with rocky and metallic meteoroids. These rocky snowballs range in diameter from a few hundred feet to several miles. This is currently the most commonly accepted model.

☐ Another prominent astronomer, Dr. Raymond Lyttleton, thinks that comets are clouds of individual meteoroids, held together by weak mutual gravitation.

More information about comet origin and composition can be found in Chapter 2. The mystery is complicated. There is still much to be learned about these strange cosmic apparitions. Velikovsky's theory is a special case in part because it has aroused controversy in the scientific community. The story of this theory, the almost cult-like reactions of its adherents and the near-hysterical opposition of its critics, is discussed in Chapter 6.

MAJOR IMPACTS

If we ask how the Solar System and the Earth were created, how they have evolved, and what comets and meteors are, then it is natural for us to wonder how our planet will ultimately be affected by these objects in the future. Will comets or meteorites put an end to the Earth or to life on Earth? It is unlikely but possible. Imagine the effects of a major comet or meteorite impact.

A large comet, striking the Earth directly, would produce a catastrophe of major proportions. As the comet neared the Earth, it would become extremely bright. The tail of the comet would fan out and cover much of the sky if it was outward bound from the Sun. The comet's head would be dominating in luminosity if it was inward bound. The sight would most certainly be terrifying. Astronomers would have long since calculated the point of impact.

As the comet approached the Earth and entered

Fig. 1-18A. A view of Jupiter, the largest of the planets, showing the Great Red Spot (courtesy of NASA).

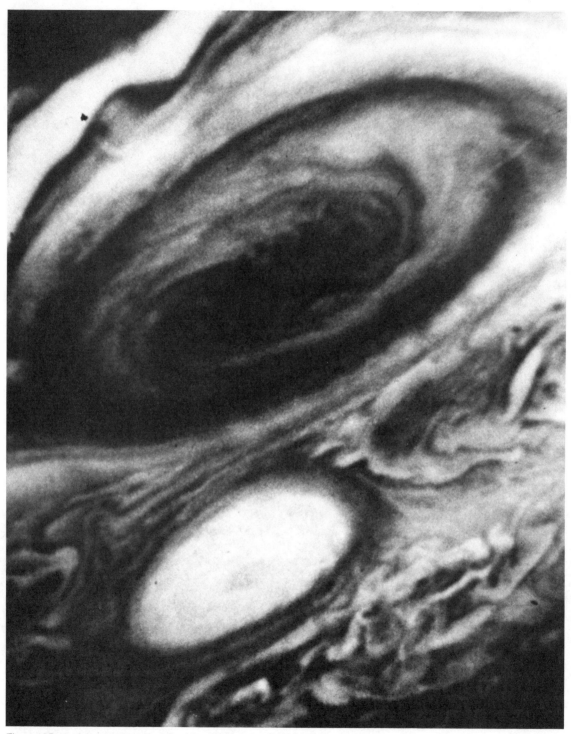

Fig. 1-18B. A closeup of the Red Spot itself (courtesy of NASA).

the atmosphere, the shock wave would level everything for a radius of hundreds of miles surrounding the point of entry. A tremendous fireball would be seen—outshining even the Sun—lasting for a few seconds. From space it would be a spectacular sight (Fig. 1-19).

If the comet struck on land, there would be tremendous earthquakes. Volcanoes would probably erupt in many parts of the world. The comet might explode in midair before striking the ground, sending meteorites in all directions like a multiple-warhead ballistic missile. Numerous impact craters, some as much as a mile in diameter, would be formed.

Small fragments of comets have hit the Earth in recent times, the most noteworthy occurring in the Tunguska region of Siberia on June 30, 1908. The explosion caused by the Tunguska Event was similar to that produced by a nuclear bomb.

Carl Sagan has pointed out a horrifying possibility. In our world today, two large countries are poised for an all-out nuclear war. Hundreds of rockets, each carrying one or more powerful, lethal weapons, are waiting to be launched from either nation toward the other. Suppose a piece of a comet strikes the Earth within the territorial boundaries of one of these countries, causing a violent fireball—complete with mushroom cloud, blast wave, and searing heat. Would that country think the other side had made an attack? Could a comet fragment cause the two rival nations to hurl themselves at each other and destroy humankind? If a whole comet struck the Earth, however, we would probably realize that it was a cosmic, and not a human, event. Astronomers would, moreover, see the approach of a complete comet.

If a whole comet landed in the ocean, the resulting tsunami (seismic seawave or "tidal wave") would propagate several times around the planet, inundating coastal plains everywhere. One or more trenches would be gouged in the bottom of the sea. There is some evidence that a trench in the Caribbean was dug out by a comet as it splashed down and impacted against the seafloor. Other trenches could be of cometary origin.

The preceding scenarios are based on the assumption that the Whipple model of comets is the correct one—that the objects are chunks of ice with meteoroids imbedded in them. What if the Lyttleton model is right and comets are actually clouds of tiny meteoroids with no solid structure? Then there would probably be no catastrophe even if a comet made a direct hit. We would simply get an incredible meteor shower, the sky set ablaze with billions of shooting stars. Meteor showers are generally believed to be caused by broken-up comets. This notion is consistent with both the Whipple and Lyttleton models.

A meteorite, comparable in size to a large comet (several miles across), would produce an even greater disaster than a comet if it struck our planet. All of the material in a meteorite would be rock and iron, and the object would therefore be more massive. The impact, if on land, might cause a magma eruption like no volcano ever witnessed in modern times. The molten rock would spill over vast tracts of land, flowing through valleys and along rivers, creating "maria" on the Earth similar to those that we see as dark areas on the Moon!

In the prehistoric eons, such cosmic disasters almost certainly took place. The collisions could have been responsible for such events as global warming and cooling, ice ages, and changes in the distribution of forests and deserts. It has even been suggested that the North Pole was once in the middle of the Pacific Ocean, the South Pole was in Africa, and that a meteorite impact caused a shift in the Earth's axis. If the impact meteorite was large enough—a small asteroid perhaps 50 miles in diameter—the orbit of our planet might even be altered slightly by a collision.

How would all of this affect humans, and life on Earth in general? That is hard to say. Many cities would be leveled by earthquakes, buried with molten lava, or washed away by waves hundreds of feet high. Fires would rage out of control through forests and across prairies and savannas. There would probably be a period of general cooling because of dust in the atmosphere. Such a major climatic change would cause a great famine because of the shortened growing seasons. The human species is remarkable in its ability to adapt to sud-

Fig. 1-19. Its tail fanning out over Africa and Europe, a comet splashes down near the island of Madagascar (Earth photograph courtesy of NASA).

den and severe changes in weather or climate, but only to a certain extent. And there is always the chance that the impact of just a small comet fragment will provoke us to annihilate ourselves.

THE END OF THE SOLAR SYSTEM

Comets or meteorites might someday be responsible for the end of humankind on Earth, but suppose we survive all attempts by such objects to smash our planet or kill us. As millions upon millions of years pass, we will probably gain technological knowledge far beyond what we have today. Even within the next century, we might develop the capability to destroy or divert a comet or asteroid in space before it can reach Earth. But let us look at the extreme long-term picture: the ultimate demise and death of our parent star.

Most astronomers believe that the Sun will eventually begin to swell in size. The Sun is now less than a hundredth of the diameter of the Earth's orbit. As the supply of hydrogen fuel begins to run out at the center of the Sun, the fusion reaction will occur in larger and larger shells around the center. This will cause the Sun to get larger, and its surface will cool off. The Earth will get hotter as the Sun fills more and more of the sky.

At first, a few billion years from now, the warming will be imperceptible. Gradually the polar ice caps will melt, and this will raise the level of the oceans. Snow-capped mountains will give up their locked-in water. Rainfall patterns will change; some forests will turn to desert while some deserts become forested. The oceans will get warmer, and this will spawn hurricanes more violent and numerous than those of today. Solar astronomers will warn civilization that the time has come to find a new home; humans will have to move to another solar system.

As the Sun grows inexorably larger, the Earth's oceans will boil, the atmosphere will be blown off into space, and all traces of life will be destroyed. Not even a single bacterium or virus will remain. The Sun might bloat so much that the Earth could be engulfed by the photosphere (Fig. 1-20). If that happens, the whole planet will burn up and cease to exist. Comets and asteroids that venture within the Earth's orbit will likewise be vaporized.

Mars will become as hot as Mercury is today. The presently frigid gas giants—Jupiter, Saturn, Uranus, and Neptune—will lose their shrouds of hydrogen, ammonia, and methane. For a time, one or two of these planets might have a climate similar to that of Earth right now. But there will not be enough time for life to evolve on such a planet because the Sun will be extremely unstable.

Our parent star will shrink following the giant stage, and in the end, it will stop shining altogether. The outer planets, comets, asteroids, and meteoroids will continue to obediently orbit a dense, black ball of matter having the size of the Earth but the mass of the Sun.

The above sequence of events, most commonly accepted as the way in which the Solar System will die, would take place over several tens of millions of years. There is another possibility; the sun might explode. If our parent star were significantly larger than it is, an explosion, or supernova, would be likely. Fortunately, the Sun does not appear to be massive enough for this to happen. But we might have made a miscalculation.

A solar explosion (supernova) would occur with practically no warning. Solar astronomers might be able to tell us that we had a few days to get off the planet and a billion miles into space before we were subjected to million-degree heat.

Suppose that the Sun goes supernova starting tomorrow morning at three o'clock. You would not notice anything right away, because it would be dark, but there would be a fantastic display of the aurora (northern lights). The dawn would seem to come early. You would turn on your favorite morning television show and be informed that cities in Europe were reporting temperatures in excess of 100 degrees Fahrenheit (40 Celsius). The sky would turn a bloody red as the Sun approached the horizon. Then your television screen would go blank as contact with the network was lost.

By midmorning, the temperature would already be over 100 degrees Fahrenheit. The Sun would shine so brilliantly that you could not even look outdoors without eye protection. The electricity would fail at ten o'clock. You might want to run to your

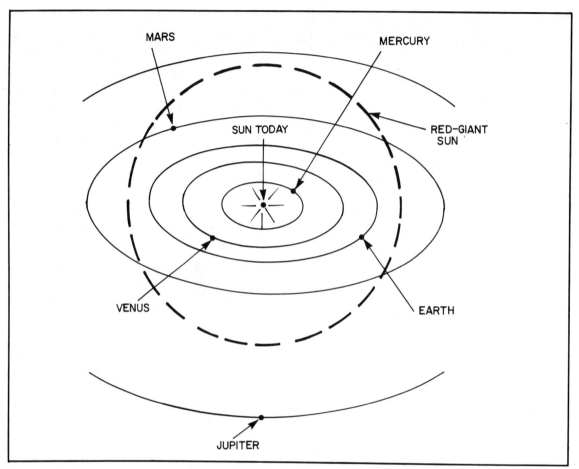

Fig. 1-20. The Sun, in the red-giant phase of its evolution, might grow to such size that Mercury, Venus, and Earth are consumed.

basement to get as much protection as you could from the mounting heat. By noon fires would start. Automobile and truck gas tanks would explode. Damaging winds would blow in the agitated atmosphere.

By evening, there would be few people left alive. The night would bring only partial darkness as the sunlight reflected off of zodiacal meteoroids and dust. The Moon would appear hundreds of times brighter than normal. The winds would increase to hurricane force and the atmosphere would escape into space. By the next day, the Earth would be as barren, charred, and desolate as Mercury. After one more day, the planet would probably break up and vaporize. Mercury and Venus would likewise be destroyed; Mars, too, might not survive. The outer planets would probably be stripped of their shrouds of gas, as a supernova lasts for up to a month. Comets in the distant halo beyond Pluto would probably flare up as though they were Sungrazers (Fig. 1-21).

Fortunately, the probability of this happening tomorrow, or even within a million years, is essentially zero. Astronomers are reasonably certain that our parent star will never have this sort of temper tantrum.

THE END OF THE UNIVERSE

Our universe is expanding, and it will continue to

Fig. 1-21. Distant comets, previously unscathed, are set aglow by a supernova sun.

expand for billions of years. What is its ultimate fate? Will the cosmos keep expanding forever or will it fall back together again? This is an important question to scientists trying to unravel the mystery of the "big picture."

The Ice Theory

Whether or not the universe will ever fall back together depends on how much matter there is. The more matter the universe contains the greater its density, and the stronger the gravitational force acting on every galaxy, star, planet, comet, asteroid, meteoroid, dust speck, and atom. According to recent estimates, our universe does not contain enough matter to pull itself back again into ylem. There appears to be only about a tenth of the matter necessary for this to happen. If our measurements are correct, the universe is speeding outward with too much force for gravitation ever to overcome. This situation is diagrammed in A of Fig. 1-22.

Every star is eventually doomed either to explode, throwing its substance back into space, or to shrink and cool off until it is a dark, cold sphere of incredible density called a black dwarf or a time-space singularity known as a black hole. From the scattered entrails of exploded stars, new stars form. Some of them explode and some of them become black dwarfs or block holes. The process continues like the natural cycle in a forest where trees sprout, grow, die, and nourish the soil for new seedlings.

How long can this cosmic cycle continue? Perhaps the answer is hundreds of billions of years. Eventually, all of the hydrogen in the universe will become heavy matter consisting of atoms too complex to support nuclear fusion. Then there will be no more stars. Without the light and heat that stars provide, the universe will grow dim and cold. Finally there will be only dead matter in a completely black, uninhabited cosmos. We might call this the ice theory of the death of our universe.

The Fire Theory

The notion that the universe will perish in darkness and cold is rather depressing. Something within us revolts against such an idea. Perhaps we are wrong about the amount of matter in the universe. It is hard to imagine how we could be off by a whole order of magnitude—a factor of 10—but it is possible. We have been wrong before about many things.

How do we rationalize such a large error so that we can formulate a theory in which the universe falls together again? When we try to figure out the density of matter in our cosmos, we must base that guess on what we see: stars, galaxies, and nebulae. We cannot include planets, asteroids, meteoroids, and comets except to guess that they contribute as much mass to other star systems as they do to our own. That's not much; it's less than 1 percent.

We do know that there is plenty of invisible matter out there in the form of gas and dust, small, dense white-dwarf stars, black dwarfs, and black holes. Could it be that visible things account for only a tenth of all the matter in the universe? Possibly! Then there is enough mass to allow gravitation to overcome the expansion, and the cosmos will be able to pull itself back together eventually (B of Fig. 1-22).

If this is indeed the case, it will be many billions of years before the atoms of the universe fall into each other (collapsing like a replay of the Big Bang run backwards). The temperature of the cosmos will rise again to unimaginable levels. All things will contract to a single point in space, and time will come to a halt.

Some astronomers think that this chain of events, which can be referred to as the fire theory, implies an oscillating universe. The cosmos repeatedly expands and contracts like a gigantic spring, passing in and out of a singularity in space and time at regular intervals.

WE ARE FINITE

Perhaps humankind will survive all of the celestial fragments hurled at the Earth by the cosmos. If so, we will have to meet the challenge of the deterioration of our parent star by moving to another solar system elsewhere in the Milky Way or in some other galaxy. When that star begins to die, we will have to move again, like a transient apartment

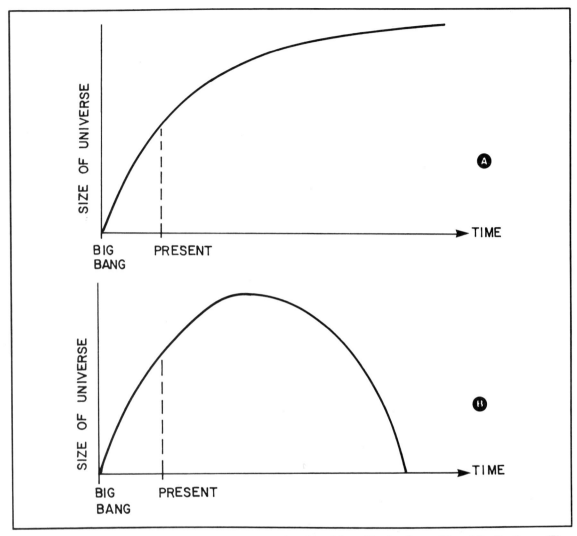

Fig. 1-22. Graphs showing the size of the universe as a function of time. The ice theory (A) and the fire theory (B).

dweller who always has some problem with his landlord. Still, humankind might survive—but not forever. In some remote future time, the clock will run out for our universe. Maybe the end will be cold, dark, and sparse. Maybe it will be hot, bright, and dense. But it will come.

We might, in an intellectual straw-grabbing attempt to rationalize that our species is immortal, argue that there are other universes into which we can dive when ours falls to pieces or burns itself up. That is pure speculation. The best answer, if we are to insist on immortality, is provided not by science but by theology.

We are finite. That lesson is taught by a finite cosmos. The lesson has been, and will again someday be, brought to us by some wandering fragment of ice, rock, or metal from outer space. The next "cosmic reminder" is due to grace our skies in late 1985 and early 1986. It will pass. It will not threaten us with any harm at all—this time—and will begin its journey back into the far reaches of space. It is called Halley's Comet.

Chapter 2

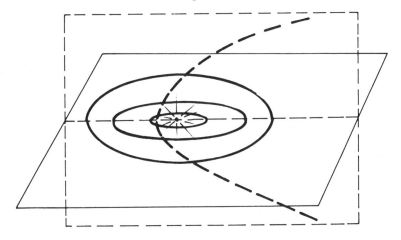

Fragments of Snow and Rock

ALMOST EVERYONE GETS TO SEE A COMET at least once during his or her lifetime. Most visible comets look like faint, fuzzy or wispy stars that are hardly distinguishable from ordinary stars. The difference becomes really noticeable only when the comet is viewed through binoculars or a small telescope. A few comets get bright enough to allow a tail to be clearly seen with the naked eye. Spectacular comets rarely appear. When they do appear their tails stand out against a moonless night sky, spanning half the celestial canopy and seeming literally suspended over the world.

The word "comet" means "hairy star", and that is an accurate description of how most comets look. When a "star" grows long, flowing tresses, it is natural for people to wonder about it. Comets have sometimes filled observers with dread.

With the possible exception of total solar eclipses, comets have scared people the most of any type of cosmic event. Today we know it is silly to be terrorized by a comet, but to the ancients, hairy stars were horribly ugly and mysterious. Surely they brought only disaster! Because bad things are always happening in this world, some disasters naturally occur during or just following the appearance of a comet. Comets make good scapegoats.

THE PARTS OF A COMET

A comet consists of a bright point or disk of light called the head, a surrounding diffuse region called the coma ("hair"), and sometimes a long streamer known as the tail. The very center of the head is called the nucleus.

For scientists, the nucleus of a comet is the most interesting part. It is still not known with certainty what cometary nuclei are made of, and it is possible that comets have highly variable structures. Perhaps some comets are like dirty snowballs while others are swarms of tiny meteoroids bound together by their mutual gravitation. The snow-and-rock model (Fig. 2-1A) is generally attributed to Fred Whipple, and the meteoroid-swarm model (Fig. 2-1B) to Raymond Lyttleton.

The tail is visually the most spectacular part

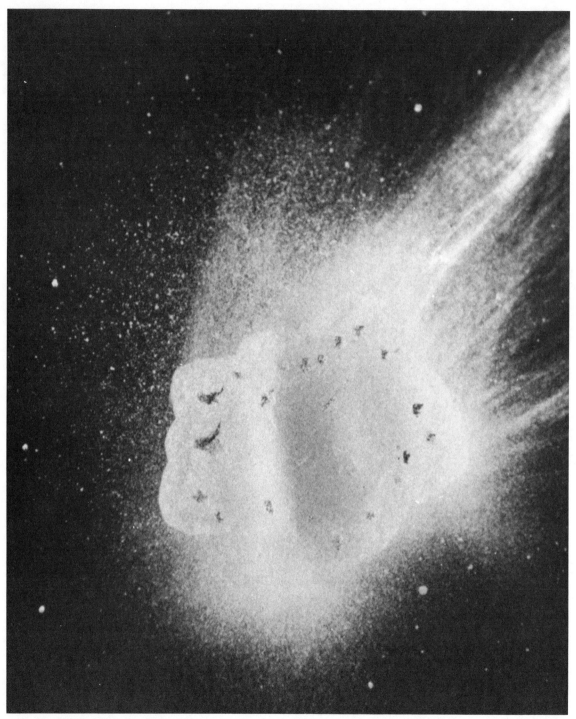

Fig. 2-1A. The two most popular models of comet nuclei are the Whipple dirty-snowball model and the Lyttleton meteoroid-swarm model (Fig. 2-1B).

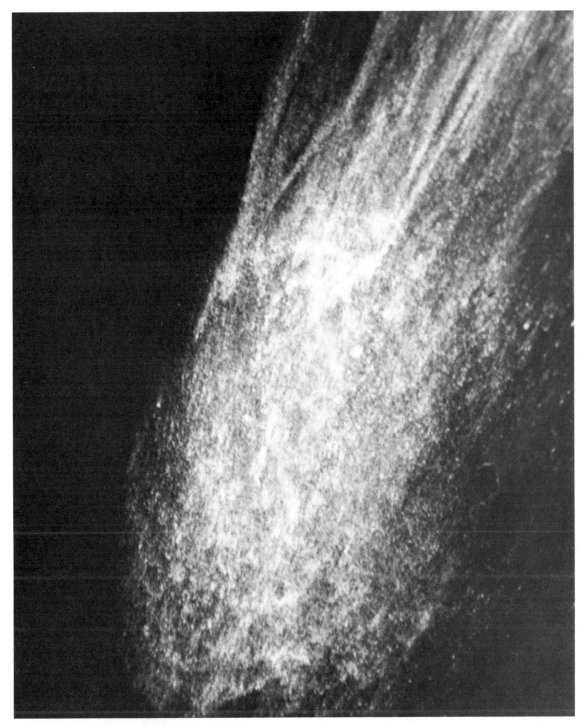

Fig. 2-1B. The Lyttleton meteoroid-swarm model.

of a comet. Tails can vary in length from practically zero (as seen from the Earth) to over a hundred million miles. When a comet is quite far from the Sun, it will not have a long tail. If the comet comes within about one astronomical unit (93 million miles) of the Sun, a significant tail is likely to form. The apparent length of the tail depends on the angle at which we see it (as well as its actual length).

Figure 2-2 shows a typical comet, including the head, coma, and part of the tail. The nucleus is obscured by the glowing coma.

EARLY THEORIES ABOUT COMETS

Some comets have tails that look like war sabers or scimitars (Fig. 2-3). For this reason, emperors of long ago were often led to believe that a comet represented a supernatural sign, indicating that the time for a war was at hand. Whether a particular ruler thought the war should be offensive or defensive seems to have been the function of a more pragmatic consideration: the strength of his military machine.

History is full of examples of wars that were affected by comet apparitions. Ironically, the comets might have been contributing factors to some battles (a self-fulfilling prophecy)! The conquerors-to-be, seeing a cosmic sword, would take advantage of it.

Probably the most well-known instance of this occurred in the eleventh century. Halley's Comet (not yet given that name) appeared in 1066 just prior to the invasion of England by the Norman French. The Anglo-Saxons thought seeing the comet meant they would be attacked. The Normans figured that the comet was telling them the time was right for crossing the Channel and invading the island. Whether or not the comet had any actual effect on the outcome of the war (the French would probably have invaded anyway), it is possible that the apparition did, by coincidence, occur at an optimal time—giving the French a little psychological advantage.

In 1456, Halley's Comet appeared as Turkish armies were attacking the city of Belgrade. Apparently the Turkish attitude toward celestial intervention in human affairs was different from the French idea four centuries before. The Turkish soldiers believed that divine forces were warning them of the dangers of war. Both sides prayed for deliverance from the evil omen so that they could get on with the battle.

Astronomers and philosophers generally were curious rather than fearful when comets appeared. The scientists were concerned not with what comets meant, but with what they actually are. Some believed that comets were of Earthly origin, representing energy or matter being thrown off into space. Aristotle was foremost among the promoters of this theory. Others, such as Hippocrates, thought that comets were interstellar objects, far from the Earth, having little effect on our planet except to make people notice. Then the Roman Empire declined and fell, the barbarians overran the civilized world, and scientific inquiry all but ceased to exist.

MORE RECENT IDEAS

As the scientific darkness and ignorance of the Middle Ages waned, astronomers—no longer fearing reprisals because of religious dogma—began to speak more freely. The old geocentric theory—the idea that the Earth is at the center of the Solar System—was abandoned. Kepler showed how the planets describe elliptical paths around the Sun. Newton discovered how gravity works (if not what it actually is). People realized that comets (like planets) are members of the Solar System, orbiting the Sun in elliptical and hyperbolic paths.

Astronomer Galileo Galilei, who lived in a time still largely influenced by the religious establishment, built a device that made objects appear many times closer than they really were. When he trained his device, called the telescope, on the skies he found that the Moon had mountains, that Jupiter had moons of its own, and that Saturn had rings. Later astronomers built improved versions of Galileo's telescope, and they also trained them on comets.

The Telescope

Telescopes do not make things look closer—only

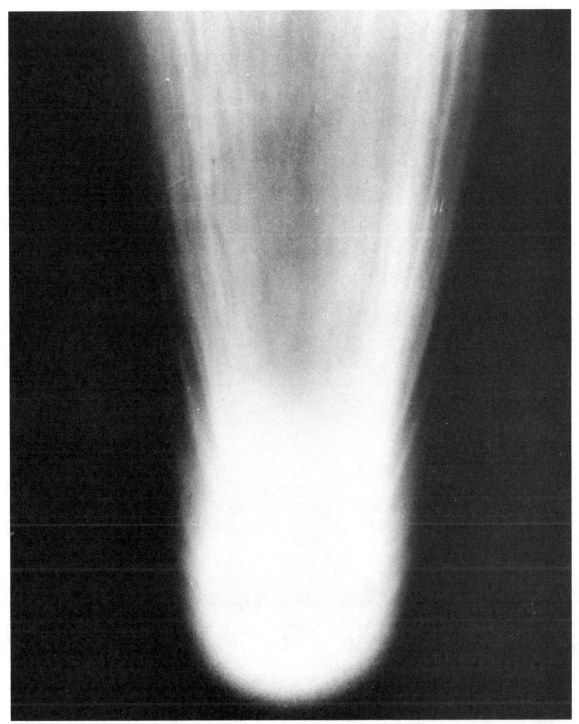

Fig. 2-2. Comet West as seen on March 5, 1976. This comet disintegrated into four pieces as it passed near the Sun (courtesy of Smithsonian Astrophysical Observatory).

Fig. 2-3. A hypothetical curved-tail comet, as it might appear over Miami, Florida.

bigger. On the scale of interplanetary and interstellar space, though, bigger is just about equivalent to closer.

Galileo's early telescope was constructed using two lenses, one with a long focal length and the other with a short focal length (A of Fig. 2-4). This is called a refracting telescope (because lenses refract light rays). The amount of magnification depends on the ratio of the focal lengths of the two lenses. If, for example, the objective lens has a focal length of 1 foot while the eyepiece has a focal length of 1 inch, the telescope will make distant objects look 12 times closer.

Refracting telescopes have certain inherent shortcomings, and this eventually led to the development of the reflecting telescope. A reflecting telescope uses a mirror instead of a lens to gather the light from space (B of Fig. 2-4). Reflecting telescopes can be made much larger than refracting telescopes because the mirror can be supported from behind. This allows astronomers to see much dimmer objects.

As you look at the sky on a dark night, your pupil dilates. It might have a diameter, after your eye has adjusted to the darkness, of about one-quarter of an inch. The light-gathering capacity of a telescope (and your eye is, in fact, a small telescope) depends on the size of the lens or mirror that picks up the light. If you double the diameter, you quadruple the effective light-gathering power. It is light-gathering capacity that is especially important for discovery and location of comets in contemporary astronomy. Until they come near the Sun comets are dim.

Magnification is important for probing into the nucleus of a comet. For a telescope, the maximum resolving power is dependent on the ability of the device to gather light. A telescope can provide about 50 to 60 magnifications per inch of objective-lens or mirror diameter. Attempts to gain more magnification give no extra resolution.

A 1-inch telescope has about 16 times the light sensitivity of your eye. A typical 4-inch telescope is 256 times as sensitive as your eye. The telescope at Palomar Mountain, which is almost 17 feet across, has approximately two-thirds of a million times the light-gathering ability of your eye on a dark night! It also has two-thirds of a million times the resolving power—or capacity to see detail—as the naked eye.

Once the telescope was in common use, comets were found more and more frequently. They

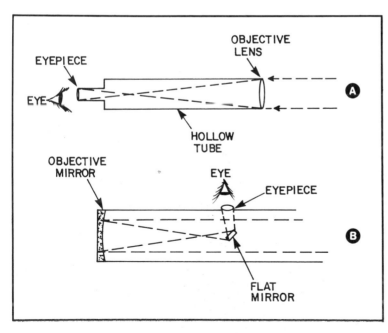

Fig. 2-4. Two common types of telescopes used for viewing astronomical objects are the refracting type (A) and the reflecting type (B).

were found to be as numerous as or perhaps more numerous than fishes in the sea. Or so we would guess because comets are still being discovered at an astonishing rate.

The Spectroscope

Most attempts to evaluate cometary nuclei with telescopes have been disappointing. Even with high magnification, a comet's head looks either like a diffuse blob or, in some cases, a tiny, starlike point of light. There are other devices that can be used to observe astronomical objects, including comets. One of the most valuable is the spectroscope. This device is employed to determine the makeup of a distant celestial object.

Visible light consists of energy at many wavelengths. The energy is propagated in the form of electric and magnetic fields at right angles with respect to each other, and having certain frequencies of oscillation. White light is made up of electromagnetic fields ranging in wavelength from about 0.00039 to 0.00075 millimeter. Each second, the fields oscillate from 400 million million (400,000,000,000,000) to almost 800 million million (800,000,000,000,000) times! The wavelength is too short to see without the aid of special instruments. (Astronomers specify wavelengths of visible light in tiny units known as Angstroms; one Angstrom is one ten-millionth of a millimeter.)

If light has just one wavelength, it appears colored. The longest wavelengths, in the range of 7500 Angstroms, look red to us, and the shortest, around 3900 Angstroms, look violet. In between are orange, yellow, green, blue, and indigo. Figure 2-5 shows the visible-light spectrum as seen by human eyes.

The various wavelengths or colors of light, although not distinguishable in white light, are nevertheless there. A prism allows us to separate the colors—obtaining a rainbow. This effect was noticed by physicists centuries ago. Curious astronomers attached prisms to telescopes (Fig. 2-6) and observed the spectra of the Sun, planets, nebulae, and stars. The familiar rainbow patterns always appeared, but something was strange about them: the spectra were not continuous. Some ob-

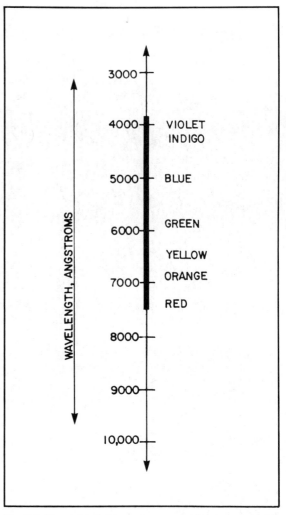

Fig. 2-5. The visible-light spectrum, according to wavelength in Angstrom units.

jects had dark lines in their spectra, and some spectra consisted of numerous bright lines. This effect was most vivid when the light was passed through a narrow slit before going through the prism.

Gases are transparent to electromagnetic energy at most wavelengths, but light at certain frequencies cannot pass. For example, oxygen is opaque to light having wavelengths of 7594 and 6867 Angstroms; sodium refuses to pass energy at 5893 Angstroms; iron is opaque at 5270 Angstroms. This effect was first noticed by a German optician named Fraunhofer as he observed the spectrum of

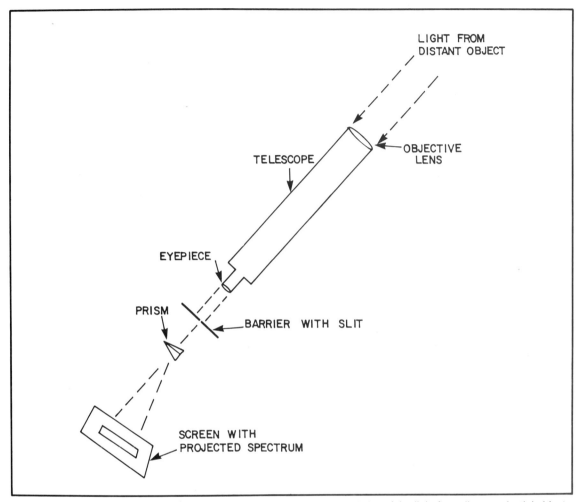

Fig. 2-6. Prisms are attached to telescopes to view the spectral characteristics of the light from distant celestial objects.

the Sun. Modern astronomers call these lines Fraunhofer lines. Some of the most well-known Fraunhofer lines are depicted by A of Fig. 2-7.

So constant are the Fraunhofer lines, and unique for each element, that scientists are able to identify chemical substances in the most distant celestial objects—even other galaxies. This is how we know that all of the matter in the universe is made up of the same elements as the Earth. There is oxygen, iron, and sodium, and all the other natural elements, all over the universe. For seekers of extraterrestrial life, is reassuring! Maybe there are civilizations like ours on other planets with the necessary ingredients for life as we know it.

When a gas is set aglow by heating or irradiation from a nearby star, the Fraunhofer lines appear in the spectrum as bright bands (B of Fig. 2-7). These emission lines have the same wavelengths, for a particular gas, as the absorption lines that appear when white light passes through. Perhaps you have observed the emission lines from various gases in high-school or college physics laboratories.

The absorption lines are produced as electrons gain energy from passing electromagnetic fields. The emission lines are electromagnetic fields emitted by electrons as they lose energy within the atoms of a substance. The details of all this are quite complicated. There is no need to get too involved with the subject here except to note that it is this phenomenon that makes spectroscopy possible. Each kind of atom—representing a particular, unique element or compound—behaves differently from atoms of any other substance. This difference can be detected with spectroscopes from millions of miles away.

COMET SPECTROSCOPY

Both the absorption and emission spectra are of interest to the astronomer who turns a spectroscope on a comet. Different parts of any particular comet show different types of spectra. Spectroscopy has led to the discovery that many comets have two different kinds of tail: a gas tail and a dust tail. The existence of gas tails indicates that comets are often made of material that vaporizes at a moderate temperature. The presence of dust tails proves that comets must contain rocklike matter. In the most spectacular comets, whose tails might extend millions of miles through space and span a sizable part of the night sky, it is primarily the dust tail that we see.

How do we know the difference between the so-called gas tail and the dust tail? The spectroscope allows astronomers to determine the makeup of comet tails. The presence of emission lines in one tail reveals that it consists of vapors excited to fluorescence by the solar radiation; hence the term gas tail. The longer tail has a spectral component identical with that of the Sun. Therefore, this tail is probably composed of particles or small rocks that reflect the light they receive. This tail could also fluoresce if the particles are vaporized by ex-

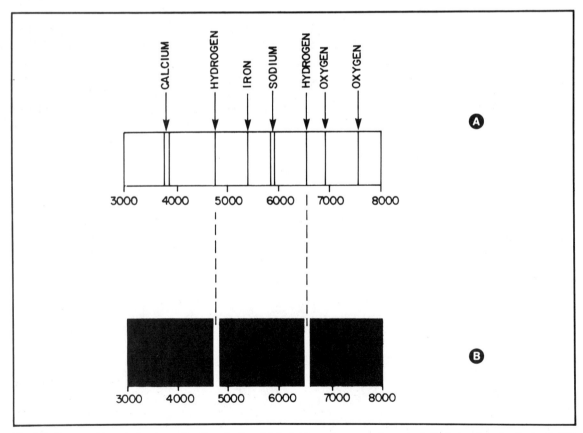

Fig. 2-7. Common Fraunhofer absorption lines for the visible-light spectrum are shown in A. Emission lines for the hydrogen molecule are shown in B.

treme radiation. It is generally thought that the particles are tiny—smaller than grains of sand on a beach—and so we call the longer comet tail a dust tail.

When a comet is far from the center of the Solar System, so that a pronounced coma and tail have not yet developed, the nucleus glows with light having the same characteristics as the Sun. This had led astronomers to suspect that cometary nuclei consist of solid particles that simply reflect light. As the comet plunges toward the Sun and heats up, the surrounding coma exhibits behavior that suggests fluorescent gases. Plain rock—or swarms of rock—alone could not account for this. The Whipple model of comets explains what we see. The Sun must cause frozen material to evaporate and glow, in much the same way as the mercury or sodium vapor glows in a street lamp. The spectra of the glowing gases indicate that comets contain hydrogen, carbon, and oxygen, along with various compounds such as methane, ammonia, and even water! In lesser quantities, metals and silicates have been identified. There is no doubt that comets are heterogeneous objects.

As a comet approaches the Sun, the comet grows in brightness. When we first see it through telescopes, hurtling toward the center of the Solar System past the orbits of the outer planets, the comet appears dim and has no tail. Only the reflected light from the far-off Sun gives us any hint that the comet is there. When the solar radiation gets intense enough—this usually occurs at a distance of 100 million to 150 million miles—the materials in the comet's core become luminous and their brightness rapidly increases.

At its nearest approach to the Sun, practically all of the comet's glow is caused by this luminescence. Even the heavier elements, such as iron, will be vaporized. Then we see a truly spectacular comet of the kind that has mystified, awed, and terrorized people for countless centuries.

HOW BIG ARE COMET NUCLEI?

The actual size of comet nuclei certainly must vary, but determining the exact diameter of a particular comet is no easy task. Figuring out the shape is even more difficult. The most powerful telescopes tell us very little, because by the time the comet comes anywhere near Earth, the glowing coma obscures the details of the nucleus itself (Fig. 2-8).

One method that has been used in an attempt to measure the diameter of nuclei is observation of comets that pass directly between the Earth and the Sun. The Sun, being so much brighter than the most brilliant comet, would render the nucleus of any comet visible as a spot or disk against the solar photosphere (provided the nucleus was large enough and was a solid body, as Whipple suggests). Nothing definite has ever been found in this way. In 1910, however, Halley's Comet did pass across the solar disk and could not be seen.

This lead astronomers to conclude that the nucleus must be smaller than 50 miles across if it is a solid body. But perhaps it is a tenuous swarm of meteoroids. Then it might be quite large, and we would not see it against the brilliant background of the Sun.

Of course, if a comet's nucleus is a solid object, it is probably opaque to the light from distant stars. If we know the speed of a comet at a certain moment, and if we know how far away it is—and the comet eclipses several stars—we can get some idea of the size of the nucleus. Rarely have comets been observed to eclipse stars. This would seem to indicate, if the Whipple model is right, that comet nuclei are small. Or perhaps Raymond Lyttleton's meteoroid-swarm model is more accurate, and most comet nuclei are nearly transparent.

Occultations have been seen in a few cases. One such event happened in 1890, but it provided little information. A single event cannot give conclusive results because the center of the nucleus might not pass directly in front of the star (A of Fig. 2-9). Numerous eclipses, however, would produce more and less occultation, allowing astronomers to get a better idea of the actual diameter (B of Fig. 2-9). Because this has never been possible due to the rarity of occultations, the nuclei of comets must generally be quite small if Whipple is correct. They could be larger according to the Lyttleton theory.

We know, at least, that comets are bigger than average-sized meteoroids. When a comet passes ex-

Fig. 2-8. The head of Comet Kohoutek as seen on January 14, 1973 (courtesy of Palomar Observatory, California Institute of Technology).

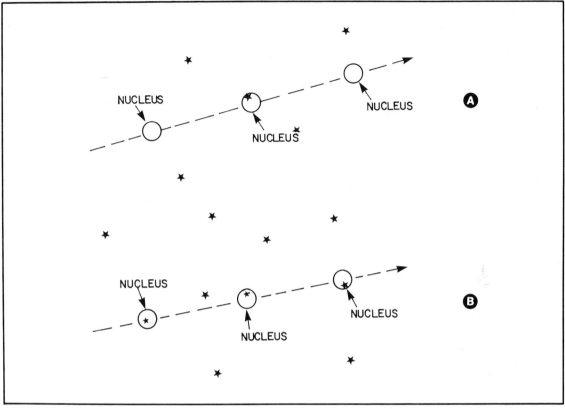

Fig. 2-9. The nucleus of a comet, passing in front of a star, might not cross it directly—giving us a false impression of the nucleus diameter (A). If the nucleus occults several stars, the discrepancies average out (B).

tremely close to the Sun, it usually survives the encounter (although it might break up into two or three smaller comets). A typical meteoroid would be entirely vaporized by the intense solar radiation within the corona of our parent star. If Lyttleton's meteoroid-swarm model is the correct one, the smaller particles would be vaporized, producing the tails, and the larger ones would gradually erode with each passage of the comet near the Sun.

How big are comet nuclei? We have only a rough idea because we still don't know precisely how comets are put together. If comets are solid bodies of snow and rock—and this is the most commonly accepted theory right now—we know that Halley's Comet is less than 50 miles across. We also know that comets are generally larger than baseballs. That doesn't narrow it down much. Perhaps when space probes visit Halley's Comet, we will at last begin to get a good idea of how substantial comet nuclei are.

THE SOLAR WIND AND COMET TAILS

The Sun produces a tremendous amount of radiant energy in the form of light and heat; to a lesser extent it produces radio waves, ultraviolet radiation (that sunburns your skin), X rays and gamma rays. All electromagnetic radiation is composed of particles called photons that travel at 186,282 miles per second through empty space. Photons don't weigh very much, but they do exert a measurable, although tiny, force on anything they strike. Radiant energy blows material off of comets, producing the coma and the tail.

The Sun also throws off subatomic particles, including electrons, protons, and helium nuclei. The subatomic particles travel much more slowly than light, but they have far more mass than the photons. Thus the subatomic particles exert more force than photons on objects near the Sun. This constant barrage of matter, streaming outward from our parent star in all directions, is called the solar wind. The solar wind is not like the Earth's atmospheric wind. The solar particles are sparse compared to the molecules in the Earth's atmosphere. In the vacuum of space, however, it takes relatively little matter to have profound effects.

As the solar wind blows gas and dust from a comet's nucleus, the tails point away from the Sun. Usually, the gas tail is almost perfectly straight and thin because the gases are ejected at fairly high speed. The dust tail is more spread out and is often curved. The curvature of the dust tail results from the fact that the particles move at relatively slow speed away from the nucleus. The nucleus constantly moves out ahead of the tail. This effect is shown in Fig. 2-10.

Many people think that comets must move "head first," with the tail lagging behind. This is true as a comet approaches the Sun. As a comet swings around and heads back out into the cold depths of space, the tail points sideways—and finally extends ahead of the head. The reason for this is clearly shown in Fig. 2-10. The solar wind always blows the tail away from the Sun.

DIFFERENT COMETS LOOK DIFFERENT

It has been said that no two snowflakes are exactly alike. The same thing holds true for comets. Whatever comets actually are, they vary in size, and probably also in shape, configuration, and composition. All of these factors affect the way a comet looks as viewed from Earth. The shape of a comet's orbit, its minimum distance from the Sun, and the relative positions of the Sun, the comet, and the Earth, also influence the appearance and the duration of the apparition.

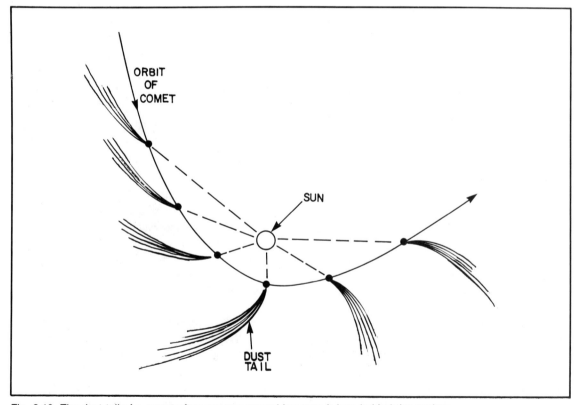

Fig. 2-10. The dust tail of a comet often appears curved because it lags behind the nucleus.

Saturn's rings, as seen from the Voyager 2 space probe. The color has been computer-enhanced to bring out the detailed structure of the rings, which move around the planet in almost perfectly circular paths. Planetary rings form from debris that cannot condense into discrete objects. It is possible that Saturn's rings formed from a comet that ventured too close to the planet and was broken up by the powerful gravitation. (NASA photograph.)

The skylab spacecraft with its meteoroid shield missing. Small meteoroids constantly bombard the Earth, and we do not even notice, but they can cause damage to spacecraft. The shield protects against smaller fragments, but there is an ever-present possibility that a larger meteoroid might strike and penetrate a space vessel. Perhaps this is what caused the near-tragedy of the Apollo 13 mission. (NASA photograph.)

The most spectacular comets are those with long tails that stand out against the evening or morning sky. A classical long-tailed comet, Ikeya-Seki appeared late in 1965. This comet was favorably situated with respect to the Earth, and it passed very close to the Sun. The dust tail extended millions of miles into space away from the Sun. The length of the dust tail, and the relatively high speed of the comet near perihelion, resulted in a noticeably curved appearance. Some comets have had tails with much more curvature, making them look like scimitars. These strange comet tails result from rapid movement of the nucleus as it passes close to the Sun.

The closer any orbiting object gets to a planet or the Sun, the faster it moves. You might have noticed this for satellites orbiting the Earth. A low-orbiting satellite takes about 90 minutes to complete one revolution around our planet. A satellite at an altitude of 22 thousand miles takes about a day. The Moon, a quarter of a million miles away, takes about a month to go once around the Earth. Planets also show this in the lengths of their years. Mercury, the innermost planet, goes once around the Sun in 88 of our days. Mars, about half again as far from the Sun as the Earth, requires almost two of our years to complete its circuit. Lonely, icy Pluto, almost 40 times as far from the Sun as our planet, takes over 200 years to make a complete trip around our parent star.

Johannes Kepler was the first to formulate a precise rule for this phenomenon. He proved that orbiting objects, whatever they are and wherever they orbit, sweep out equal areas in equal amounts of time (Fig. 2-11). This doesn't matter much for planets, whose orbits are almost perfect circles, but it is significant in the case of a comet. As one of these maverick objects comes into the Solar System, it moves slowly at first, then faster and faster, until—as it nears and passes the Sun—it moves at extreme speed, perhaps over a million miles per hour. Then the comet slows down again as it departs. Eventually, its speed may be less than that at which you drive your car down the freeway.

Because a comet is moving the fastest while its tail is oriented sideways to the direction of movement, the tail drags far behind the head. This happens as a comet moves at high speed around the

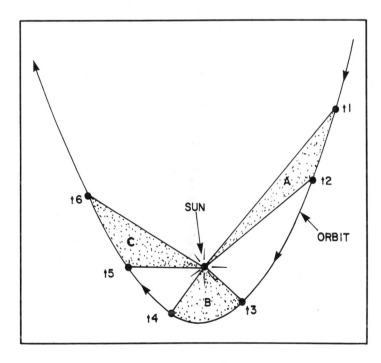

Fig. 2-11. Any object orbiting the Sun sweeps out equal areas in equal times. Time intervals t2-t1, t4-t3 and t6-t5 are all equal, and thus the area A, B, and C are all the same.

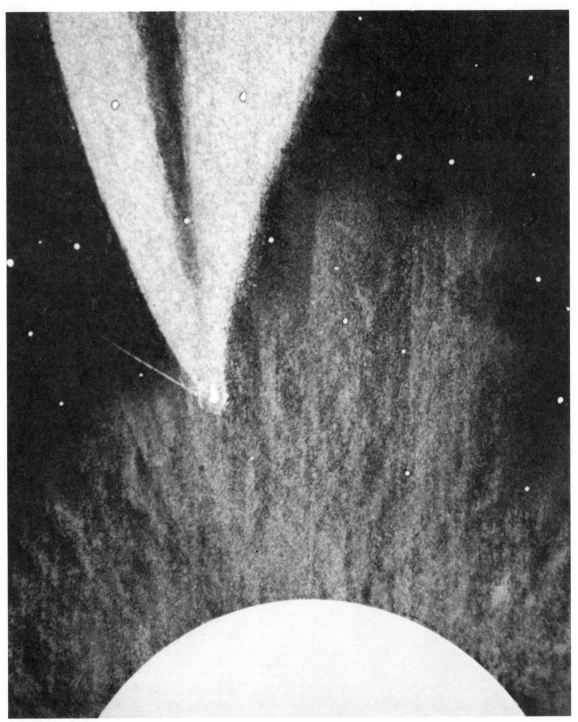

Fig. 2-12. Artist's rendition of a sungrazer comet as it passes through the Sun's corona. Sungrazer comets generally develop highly curved dust tails.

Sun at perihelion. The closer the comet gets to the Sun, the more pronounced the effect. Comets that actually enter the corona of our parent star are called Sungrazers. Such comets move with extreme speed, and they often have vividly curved tails (Fig. 2-12).

A comet could have a tail that looks straight to us on Earth, even though the tail is really curved. The appearance of a comet's tail depends on the orientation of the object's orbit. Comet orbits are almost always very elongated, but they can exist in any plane (Fig. 2-13). We might see a comet as it moves toward us as it, for example, swings over the north pole of the Sun. Or we might see a comet as it hurtles under the south pole of the Sun, sideways with respect to the Earth. Both comets might have curved tails, but we would see the curvature only in the second case.

Suppose a comet is discovered tomorrow as it falls slowly in toward the center of the Solar System past the Asteroid Belt. How do we know what it will look like—or if we will be able to see it at all—as it comes near the Sun? We can calculate its orbit on the basis of a few observations made at intervals of several hours or days. Thus we can tell just how close to the Sun the comet will pass. We will know how near it will come to our planet, and how the comet will be positioned in relation to the Sun and the Earth.

Assuming that the comet is of average size, we might be able to say with a reasonable degree of assurance that the tail will be visible or that it might be spectacular. But we're not always right. This kind of thing happened in 1972 and 1973 as Comet Kohoutek was found and its course was plotted. The comet came and went almost unnoticed even though it was at first heralded as a potentially spectacular comet. Comet Kohoutek did show up fairly well on photographs (Figs. 2-14 and 2-15). For naked-eye observers it was a disappointment.

Comets that do not come especially close to the Sun—and many do not—often go unnoticed by cas-

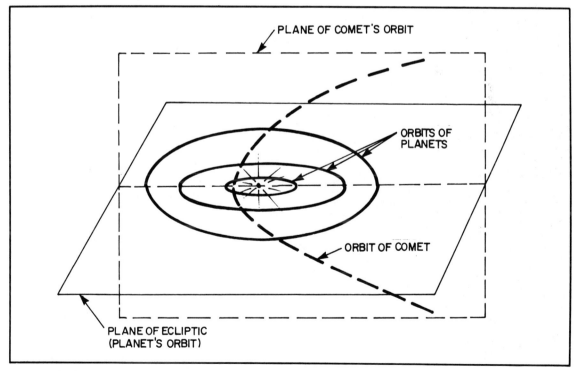

Fig. 2-13. The orbit of a comet is not necessarily in the same plane as the orbits of the planets. In this example, the comet orbits in a plane tilted almost 90 degrees with respect to the Solar System.

Fig. 2-14. The Comet Kohoutek of 1973-74. The entire comet is shown here and the head is shown in Fig. 2-15 B. This comet was typical in terms of brightness, but its behavior took astronomers somewhat by surprise (courtesy of Palomar Observatory, California Institute of Technology).

ual comet watchers. In many instances, comet tails do not develop. Some comets have been pulled by the planets into nearly circular orbits at various distances from the Sun. In this way, some spectacular comets have been turned into insignificant whispers of their former selves.

HOW BRIGHT ARE COMETS?

Comets vary in luminosity as much as the stars and planets. In order to define how bright a celestial object appears, scientists use a numerical scale by assigning a particular object a number called the magnitude of brightness. The first astronomer to use this system was Ptolemy of Alexandria, about 20 centuries ago. We credit Ptolemy with the interesting explanation of retrograde planet orbits consistent with the theory that the Earth is the center of the universe. Of course, we now know that this theory is incorrect. Nevertheless, his system of magnitudes has proven useful.

Ptolemy observed the brightest stars, such as Spica and Pollux, and assigned to them the first magnitude. Stars noticeably less bright were considered to be of the second magnitude; still less brilliant stars were called third-magnitude stars. Ptolemy got down to the sixth magnitude for stars which appeared so dim they could scarcely be seen at all.

Modern astronomers have refined the Ptolemaic system, making it mathematically precise. A change in brightness of one magnitude represents a difference of 2 1/2 times. Thus a star of the first magnitude is 2 1/2 times as bright as one of the second magnitude, and a third-magnitude star is 2 1/2 times as brilliant as a star of the fourth magnitude. These differences can be measured with sensitive light meters. Photography can also be used. With the help of giant telescopes, extremely dim stars—fainter than the fifteenth magnitude—can be seen. Magnitudes can be measured not only to the whole number, but to the tenth, hundredth, or even thousandth part.

It turns out that, according to the modern definitions, some stars and planets have magnitudes of less than one. This strange accident (this kind of thing is actually quite common in the realm of

Fig. 2-15. The Comet Kohoutek (courtesy of Palomar Observatory, California Institute of Technology).

science) has resulted in magnitudes of less than one or even less than zero. For example, Sirius, a star in the constellation Canis Major, has a visual magnitude of -1.43. Venus, usually the brightest object in the sky except for the Sun and Moon, is still brighter at certain times. Comets range in brilliance up to about the apparent luminosity of Venus at its brightest.

Stars shine by their own light, and although they might appear to be quite dim, they are also extremely distant. Actually, stars are bright—much more brilliant than the brightest comet—insofar as absolute luminosity is concerned. A close, dim object might look much brighter than a far-off, brilliant object in space. Planets, shining by reflected sunlight, are also much more brilliant than comets. Nevertheless, it does not take a very luminous comet to create a spectacular sight in the sky on a moonless night away from city lights!

The tail of a comet almost always appears less bright than the head or coma. This is because the material in the tail is far more tenuous, and the glow is spread out over a larger region of the sky. You have no doubt observed this effect with diffused light in a room, as compared with the harsh glare of illumination from unshaded incandescent bulbs.

A few comets get bright enough so that they are visible even in daylight. These events are so rare, that they happen only once every few centuries. Usually, only the head and perhaps a small part of the tail are visible while the Sun is above the horizon.

COMETS CHANGE APPEARANCES AS THEY MOVE

The way a comet looks changes as the nucleus progresses around the Sun. The gas tail might look long, short, or invisible. Sometimes the gas tail seems to point toward the Sun because of its relative angle with respect to the Earth, but this is rare. The dust tail of a comet might appear straight, curved, thin or broad. Some comets develop several tails and then appear to have just one at a later time. Such a comet apparition took place in 1744 when six tails were seen as the so-called Great Comet passed close to the Sun. (There were a lot of "Great Comets" before standard nomenclature was introduced.) The Great Comet of 1861 also had a broad, complex tail. The appearance of the dust tail or tails depends on how we view the comet. The nearer a comet comes to a line directly between the Earth and Sun, the broader and more diffuse its dust tail gets.

Suppose a comet moves in a path similar to that shown in Fig. 2-16. We observe it at four different times, representing four different positions of its orbit. The Earth also moves as the comet goes by. As shown in A of Fig. 2-16, the comet is somewhere between the orbits of the Earth and Mars, and it is moving almost directly toward us. At Point B, the comet is near the orbit of our planet, and it is gaining speed. At Point C, the comet is crossing the orbit of Venus, and it is approaching a line between the Earth and the Sun. At Point D, the comet is almost exactly between the Earth and the Sun, and it is just past its perihelion. For simplicity, we can assume that the orbits of the Earth and the comet are in the same plane.

Figures 2-17A, 2-17B, 2-17C, and 2-17D show how the comet might look to us at each of these different times. When the comet is at point A as shown in Figure 2-16, the dust tail looks short and straight. It is pointing more or less away from us,

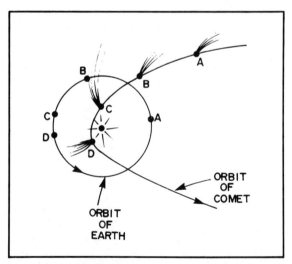

Fig. 2-16. Hypothetical comet orbit with respect to the orbit of the Earth, showing positions of the Earth and the comet at four different times A, B, C, and D.

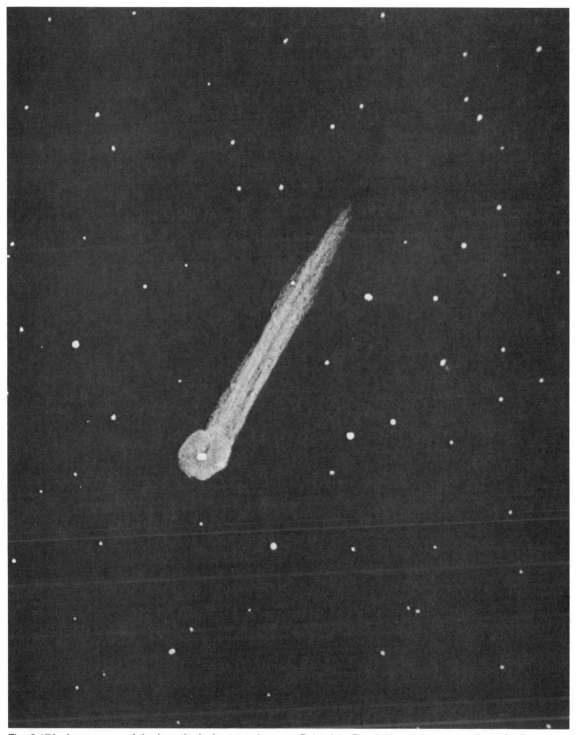

Fig. 2-17A. Appearance of the hypothetical comet shown at Point A in Fig. 2-16 at time as seen from the Earth.

Fig. 2-17B. Point B.

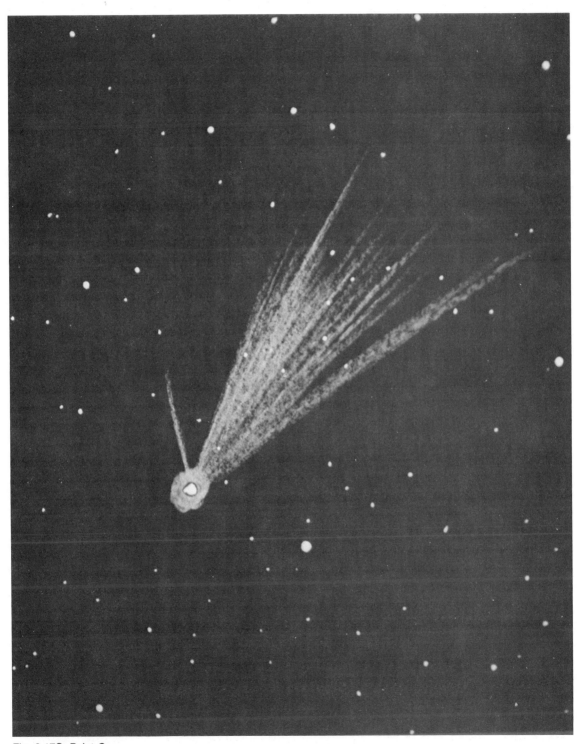

Fig. 2-17C. Point C.

Fig. 2-17D. Point D.

and the comet is moving rather slowly and is still quite far from the Sun. The gas tail is invisible, blending in with the dust tail. At B, we see the dust and gas tails from a better vantage point. Both tails look longer in proportion to their actual lengths. The tails are actually longer because the nucleus is closer to the Sun.

The slight curvature of the dust tail separates it visually from the gas tail. The curvature results from the high speed of the nucleus. At Point C, the tails are turning toward us, and this makes the dust tail appear fanned-out and the gas tail even more separated from the dust tail. At Point D, the nucleus is almost directly between the Earth and the Sun. The dust tail seems separated into several constituents. The gas tail is still identifiable as a thin, straight, needlelike white spike. By this time, we see the comet's head only just before sunrise or after sunset. The tail is seen in total darkness, however, projecting eerily above the horizon into the background of stars.

DIFFERENT APPARITIONS WILL LOOK DIFFERENT

A periodic comet will appear different from one apparition to the next in most cases. Sometimes the difference is very small, but it is often quite significant. For example, a certain comet might be best seen from the Southern Hemisphere, at one passage, and from the Northern Hemisphere during the next passage. A given comet might appear very bright—with a long, spectacular dust tail at one apparition—but hardly be seen at all during the next visit. This effect occurs for two reasons.

First, the relative orientation of the comet's path, with respect to the orbit of our planet, is rarely the same for two passages in a row. Sometimes the comet passes near the Earth; then it will look bright. On other occasions it passes on the opposite side of the Sun from the Earth; then it never gets very spectacular. This is shown in A and B of Fig. 2-18. As shown in A, the comet happens to pass close to the Earth, and it appears bright with a large tail. As shown in B, the comet reaches perihelion opposite the Earth, and might not ever become visible to the naked eye.

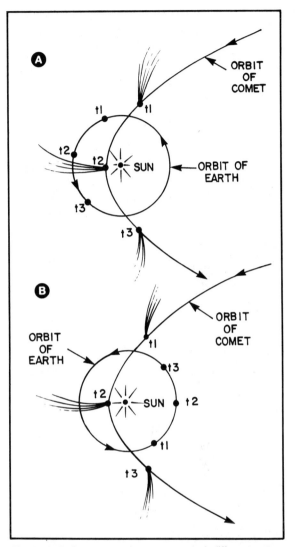

Fig. 2-18. A given comet may appear much different on two successive apparitions. At A, the comet passes on the near side of the Sun; at B, it is on the far side of the Sun. Positions of the Earth and Sun are shown for three moments t1, t2 and t3.

In 1910, Halley's Comet was especially bright because it passed near the Earth. But in the 1985-86 apparition, this comet will be less spectacular because it will never get very close. Unfortunately for the majority of humankind who live in the Northern Hemisphere, this comet will again, as in 1910, put on its best show south of the equator.

Another reason why a given comet might

change appearance from one passage to the next is simpler: some comets deteriorate. We know that comets contain only a certain amount of material, and that the tails consist of this matter blown off by the radiation from the Sun. Each time a comet passes the Sun it loses some of its mass. We would expect that a given comet would get less and less bright with each perihelion until, finally, it disintegrates altogether. Some comets have done just that: they have passed behind the Sun and have not reappeared. Other comets have broken into two (or more identifiable) pieces from one apparition to the next.

Not all comets seem to die down as they appear and reappear over centuries of time. Halley's Comet has been seen for over two thousand years. It is still a relatively bright comet. Perhaps comet nuclei somehow replenish their store of matter while they are far from the Sun in the cold, outer fringes of the Solar System. Space is not empty. There are traces of dust and various gases, mainly hydrogen. Could it be that the weak gravitation attracts dust particles and additional gas molecules each time the comet swings out on its long, slow trip into cold space? Or might interplanetary gases condense on the cold comet nuclei, like water droplets on a glass of iced tea? These questions remain unanswered.

FLARE-UPS AND BURNOUTS

Whenever a comet is discovered, astronomers begin to observe it with constant curiosity. The first questions they ask are, "How bright will it be? What is its orbit? How close will it come to the Earth and the Sun? What should we tell the general population to expect?"

Comets are always dim, fuzzy objects when they are first detected plunging inward past the outer planets and the Asteroid Belt. On the average, they gradually grow brighter. Sometimes a comet gives us a surprise. The art of predicting the future behavior of a new comet is fraught with uncertainty. Rapid changes in brightness, either increasing or decreasing, are common. Why would that sort of thing happen?

There are several various theories that have been put forth in an attempt to explain sudden changes in the brightness of a comet. Astronomers routinely graph the absolute (or actual) brilliance of comets as functions of their distances from the Sun. On the average, comets get brighter as they draw near the Sun and dimmer as they get farther away. The absolute brightness is shown on the graph along with the distance from the Sun in astronomical units. An example of such a graph, showing three different hypothetical comets, is shown in Fig. 2-19. In this graph, time progresses from left to right along the horizontal axis (although at a nonlinear rate because of the eccentricity of the orbit), and brightness is plotted along the vertical axis.

Two of the comets in the figure show regular, predictable changes in brilliance. Most comets are more or more less like this—giving us no surprises. The third comet clearly flares up and then rapidly dies out. The curve is not smooth like the other two; it contains an obvious irregularity.

If the dirty-snowball (Whipple) comet model is correct, we can explain the flare-up on the basis of nonuniformity in the structure of the comet. As a child, perhaps you made dirty snowballs in the winter. You know that the particles of dirt or sand or rock are never homogeneously distributed throughout the snowball. Instead, the dirt or sand or rock is packed in clumps, and might be concentrated almost entirely in one part of the snowball. It is probably the same with comets. As a dirty snowball melts, it often does so at a variable rate, depending on the ratio of snow to dirt exposed at any given moment as the object shrinks. Comets of this kind would behave in the same way. The faster the "melting" (actually evaporation) process the brighter the comet would appear.

Another possible cause of a comet flare-up is the occurrence of a solar flare. Especially during times of maximum sunspot activity—it reaches a peak approximately once every eleven years—sudden disturbances erupt on the surface of the Sun. These so-called flares can be seen from Earth through telescopes, and later detected by means of radio equipment.

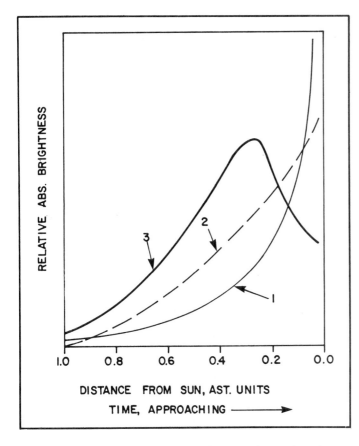

Fig. 2-19. Brightness-versus-distance curves for three hypothetical comets. Comets 1 and 2 are typical. Comet 3 shows a sudden flare-up followed by a rapid burnout.

Solar flares throw off tremendous numbers of subatomic particles that are attracted to the magnetic poles of the Earth, producing the auroral displays familiar to inhabitants of the higher latitudes. These particles might also be expected to have a profound effect on the nucleus of a comet near the Sun (Fig. 2-20). The greatly increased solar-wind pressure would generate an acceleration in the process of evaporation of the ices, and a generally more rapid breakdown of the whole nucleus. We would expect to see the coma expand and grow brighter, and the tails lengthen and become more prominent. The change could take place within hours or even minutes. And that is exactly what does happen in some comet flare-ups. Following a solar flare, the comet would be expected to get dimmer again; abrupt diminutions are often observed.

Still another potential reason for a comet flare-up is the passage of the nucleus through a region of space where there is an unusually high concentration of meteoroids. Continually peppered by small meteoroids at high speed, a comet's nucleus would become temporarily more irregular. The resulting greater surface area would allow faster evaporation and consequently greater brilliance for a time (Fig. 2-21).

If a comet were to smash into a moderate-sized or large meteoroid, the whole nucleus might break into two or more pieces. Such cataclysms would be rare, but events of this kind have been observed. As early as 1846, a comet's nucleus was seen to suddenly and unexpectedly split in two. Comet Ikeya-Seki of 1965 broke into two constituent nuclei shortly after passing the Sun. In 1976, the nucleus of comet West split into three and then four separate, identifiable parts.

The observation of comet nuclei splitting in half, or even in thirds or quarters, lends support to

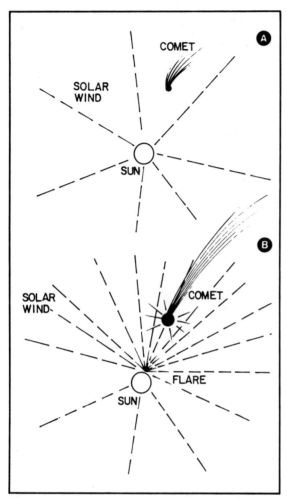

Fig. 2-20. A solar flare can cause a comet to suddenly brighten. At A, a comet approaches the "quiet" Sun. Then a solar flare occurs, throwing out subatomic particles that increase the evaporation rate of ices in the comet (B).

Sometimes comets undergo rapid and unexpected diminution in brilliance. One possible cause of this might be the breakup of the nucleus, following the complete evaporation of the icy matter, into a meteoroid swarm. Another reason might be that the comet has a rocky core surrounded by ice. When the ice burns off, only the asteroidlike center is left, and the coma and tail largely disappear (Fig. 2-22).

ROTATION OF COMET NUCLEI

Astronomers have observed comets not only for sudden changes in brightness, but for slow, regular fluctuations that would indicate rotation and irregularity of the nucleus. These observations are normally carried out while the comet is still far from the Sun (so that the comet shines only by reflected sunlight). When the comet reaches the inner planets, the glow of the coma obscures the reflected sunlight, making the observations difficult or impossible to interpret.

What have we seen? Is there any proof that comet nuclei rotate on some constant axis? Although some telescopic observations have appeared positive, indicating a rotating nucleus, the results are somewhat inconclusive. But we would expect that comets—like practically all other celestial objects such as stars, planets, moons and asteroids—tumble through space. If we find that comet nuclei never tumble, it will come as a surprise! The statistics-oriented astronomer might do well to assume that most comet nuclei do tumble. There is some good evidence, based on changes in comets' orbits, that supports this probabilistic "armchair argument."

The rotation of a comet's nucleus could perhaps affect its orbit around the Sun while the comet is near perihelion. This possibility was first suggested by Whipple in 1974. It is reasonable to suppose that most of the evaporation from a comet's nucleus takes place on the side of the nucleus facing the Sun because that is where the solar radiation strikes the nucleus. This might produce what Whipple terms a jet reaction, pushing the comet—perhaps with enough force to change its orbit slightly from that of a solid, nonevaporating object.

Whipple's dirty-snowball model. A swarm of meteoroid particles, held together by mutual gravitation, should not be affected in this way. Most astronomers today think that most (or all) comets are of the dirty-snowball type. Until many comets are investigated at close range by space probes, we will not know for certain whether or not there might be some comets of the meteoroid-swarm type. There is strong evidence, as we shall shortly see, that comets eventually lose all their icy matter, leaving diffuse clouds of meteoroids, producing meteor showers at yearly intervals.

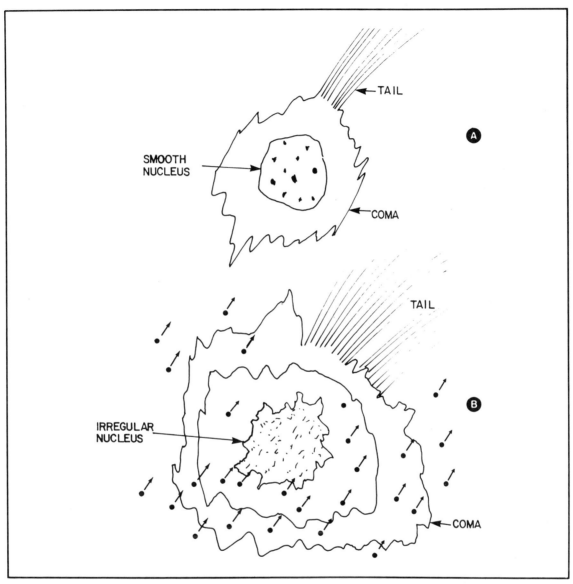

Fig. 2-21. A cloud of meteoroids might cause a comet to flare up. At A, a hypothetical comet approaches the Sun, and then passes through a swarm of meteoroids that erode its surface and cause an increase in brightness (B).

The effect of the jet reaction might be to increase the perihelion or to decrease it. The nature of the orbital change would depend on whether the comet was rotating, and also on the orientation and speed of the rotation. The period of revolution around the Sun also would thus be altered. Acceleration would enlarge the orbit and make the period longer; deceleration would cause the orbit to shrink, making the period shorter.

Figure 2-23 illustrates four possible situations for comet rotation that might have an effect on the orbit. As shown in A of Fig. 2-23, there is no rotation at all, or else it is so negligible that it has no effect whatsoever. In this case, the jet-reaction force would be outward from the Sun at all times, reaching a maximum at perihelion. This would re-

67

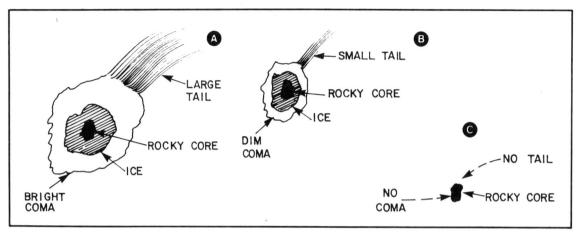

Fig. 2-22. A comet with a rocky core, surrounded by ices, would appear to burn out if the ice were depleted. At A, there is much ice; less ice at B and none at all at C. When the ice is all gone, the comet becomes a meteoroid or asteroid, shining only by reflected sunlight.

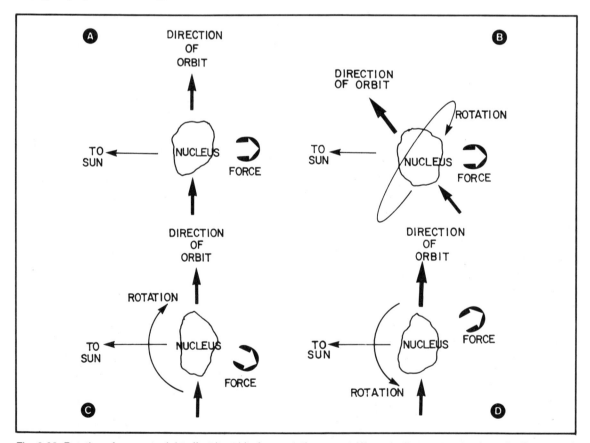

Fig. 2-23. Rotation of a comet might affect its orbit. A nonrotating comet (A) gradually moves away from the Sun at each perihelion. At B, the rotation is nearly perpendicular to the plane of the comet's orbit, and this causes a gradual change in the plane of the orbit. At C, retrograde rotation slows the comet down. At D, forward rotation speeds up the comet.

sult in a gradual increase in the perihelion distance from revolution to revolution.

As shown in B of Fig. 2-23, the comet's nucleus rotates at right angles to the plane of the orbit. The rotation causes the jet reaction to be displaced from the plane of the orbit, and the result is to push the comet's orbit to a different plane from revolution to revolution. As shown in C of Fig. 2-23, the rotation is retrograde. That is contrary to the direction of the comet's orbit, but in or near the same plane. This results in a gradual deceleration and a steadily shrinking orbit; the period becomes shorter.

As shown in D of Fig. 2-23 the rotation is in or near the plane of the comet's orbit, and in the same direction. Thus the comet accelerates and the orbit gets larger and the period longer.

It is most likely that comets tumble in all imaginable different orientations and at innumerable different rates. The effects must depend on many different factors that include the general shape of the orbit, the perihelion distance, the orientation and speed of rotation, and the size and composition of the nucleus. Predicting the extent of the effects is beyond our present capability. We can draw one conclusion: unexplainable changes in the orbit of a comet, not attributable to the effects of planets' gravitational fields, indicates (according to Whipple) that the nucleus is probably rotating. Such anomalies have been observed. Despite high-precision equipment and computers for calculating the positions and perihelions of comets, the nuclei do not always behave exactly according to the forecasts. Sometimes a comet seems to have speeded up, slowed down, or passed perihelion at a slightly different distance than expected. When space probes are sent to observe the nucleus of Halley's Comet at close range, this is just one of the mysteries that perhaps will be resolved about that particular comet.

THE EFFECTS OF THE OUTER PLANETS

On numerous occasions, a comet, plummeting toward the Sun from deep space, is deflected by the gravitational field of a large planet. Jupiter, especially, can cause this because it is the most massive planet and therefore has the greatest gravitational influence. Because they are quite massive, Saturn, Uranus, and Neptune can also affect the path of a comet. These "gas-giant" planets all lie in parts of the Solar System where comets move slowly; this magnifies their potential effect on the motion of a passing comet. Pluto might influence comets once in a while, but this planet has relatively low mass and a correspondingly weak gravity.

It is not yet resolved whether some comets originate from outside the Solar System. It is certain that many comets have aphelia (maximum distances from the Sun) well beyond the orbital radius of the outermost known planet (Pluto). Such comets would have orbital periods of hundreds, thousands, or even millions of years. A close encounter with a major outer planet, however, can radically change the orbit of a comet so that it acquires an aphelion close to the planet and, therefore, a much shorter orbital period.

A given planet, over the course of many millennia, captures numerous comets in this way. All the comets would have aphelia close to the orbital radius of the planet. Such a set of comets is called a family for that planet. Jupiter has a fairly large known family of comets. Saturn also is known to have a comet family, but that family is smaller because Saturn has a weaker gravitational effect than Jupiter. Uranus and Neptune must also have comet families, and Pluto, feeble as its gravity is, probably has a comet family, too.

Figure 1-11 illustrates how a major planet "catches" a comet, bringing the nucleus permanently within the confines of the Solar System. The comet must pass close to the planet on a certain revolution. This is most likely to happen if the comet's original orbital plane lies near the orbital phase of the planet. Eventually, if this is the case, the comet and the planet will come close to each other. Statistically, given enough time, this is inevitable. Then the comet will swing around because of the gravitational pull of the planet, and its orbit will change. The aphelion will be brought closer to the Sun (Almost always near the orbital radius of the planet.) The perihelion will be pushed correspondingly farther away from the Sun. (According to Kepler's first law, which states that objects must orbit the Sun in ellipses with the Sun at one focus.) A hypothetical comet orbit prior to capture by Jupiter is shown in A of Fig. 2-24. A probable orbit after the event is shown in B of Fig. 2-24.

It is possible for a comet to pass so close to a planet that the comet is captured permanently by the planet (becoming a moon of the planet). Jupiter might have dozens or even hundreds of comet moons. Saturn, Uranus, Neptune, and even Pluto probably have some. It is possible that the beautiful rings of Saturn, and the less prominent rings of Jupiter and Uranus, were once comets that got caught and broken up by the gravity of those planets. Astronomers have discovered that the rings of these planets appear to consist of ice. This is consistent with the popular dirty-snowball theory put forth by Whipple.

Not only the outer planets, but the inner ones too, probably have comet families. There are, for example, probably numerous comets in a family under the gravitational influence of our planet. Some astronomers have suggested that the Earth might even have a ring—perhaps made up of comet fragments, similar to the rings of Jupiter, Saturn and Uranus—but too tenuous for us to detect.

HOW LONG DO COMETS LAST?

Comets, like all things in the universe, cannot last forever. We have seen instances in which comets have disintegrated as they passed near our parent star. Comets sometimes seem to grow dimmer and dimmer with each revolution. It is reasonable to suppose that the material in a comet gradually depletes after hundreds, thousands, or millions of trips around the Sun.

The longevity of a particular comet depends on several factors. The more massive comets would live longer than the less massive ones. More dense comets would survive longer than the less dense ones. Comets with extremely small perihelia (sungrazers) would deteriorate more quickly than those that stay far from the Sun. Short-period comets, attaining perihelion relatively often, would die

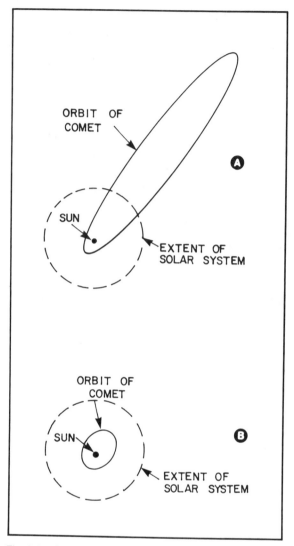

Fig. 2-24. Orbit of a hypothetical comet before (A) and after (B) capture by Jupiter.

sooner than those that have extremely long periods. Of course, we are considering these influences one by one on an "all-other-things-being-equal" basis. Everything gets more complicated when we realize that all other things are never equal!

Halley's Comet, with an orbital period of about 76 years and a perihelion of about 55 million miles, has been seen for more than two thousand years. It is almost certain that it appeared many times before the earliest recorded perihelion in the third century B.C. It is difficult to say whether its absolute brilliance has changed since then, largely because of the different orientations of its orbit with respect to the Earth on successive apparitions. As of 1910, Halley's Comet was still a relatively bright comet. It will probably remain spectacular for centuries to come.

The comet nuclei in orbits far from the Sun, never getting closer than a few astronomical units to our parent star, probably will exist until the Sun bloats into the red-giant phase billions of years from now. If there are comets wandering through interstellar space—and some astronomers think there are—many of them might last until the end of the universe, never getting the chance to pass near a star and flare up.

A few comets have a more significant fate. While the inhabitants of some distant, unknown planet look on, these fortunate comets will flare up. Perhaps the witnesses will react with emotion such as humankind has always shown toward these transitory, ghostly objects.

Chapter 3

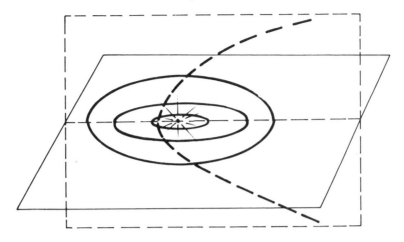

Halley's Comet and Other Famous Comets

A TRULY SPECTACULAR COMET IS A SIGHT that people never forget. The most remarkable comets have been mentioned in historical writings, and thus have obtained a certain kind of immortality in human experience. Documentation of long-ago comets has helped modern scientists to catalog returning (periodic) objects—differentiating them from one-time or "maverick" ones.

EARLY COMET APPARITIONS

In ancient times, when only the naked eye was available for observation of the heavens, people had no idea of how far away or how large comets actually were. It was generally thought that comets were some sort of atmospheric phenomenon; after all, no one knew that the atmosphere did not extend indefinitely into space! Because comet tails, when at their brightest, point upward from the horizon at a fairly steep angle, and because the objects look rather firelike, some ancient scientists suggested that comets were manifestations of energy spewing out from the Earth. What this energy might be or why it should be ejected in such a form was not known; it seems the mystery was not addressed.

An especially brilliant comet appeared in the year 240 B.C. as the Roman Empire was approaching the zenith of its power. The comet's appearance, transitory nature, and path across the sky were described in writing. Modern astronomers have extrapolated on the basis of Halley's Comet appearing at intervals of about 76 years—the last apparition having been in 1910—and have concluded that the object seen in 240 B.C. was this same comet.

We cannot be certain if this is really true because we do not know positively that a maverick comet didn't appear at about the same time as Halley's Comet would have been scheduled to appear. Astronomers are reasonably confident that it was indeed an apparition of Halley's Comet.

Looking further back in time, imagine the reactions of prehistoric men to brilliant comets! Too busy gathering food to bother much with the mysteries of the heavens, these people probably theorized that the great domelike canopy overhead was changeless. Certainly, many primitives went through their lives without ever seeing a major cosmic event, except for an occasional lunar eclipse. One evening or morning, a tenuous, thin, fuzzy patch of light would be seen near the horizon. Night after night it would get brighter. Finally it might rival Venus or even the crescent Moon as a dominating feature of the sky. It is not hard to imagine that it must have looked like a gigantic club, ready to smash the world at any moment!

In still more remote times past, it is possible that one or more comets had a great impact on life on Earth, the repercussions of which are still being felt today. The most commonly propounded theory is that the nucleus of a large comet struck the Earth, resulting in the extinction of the dinosaurs in a span of time like the blink of an eye on a cosmic scale. Some evidence has recently been found indicating that a huge number of the prehistoric reptiles perished within weeks, months, or days—all in the same part of Africa.

If this did happen, a comet impact could well have been responsible. The sky would darken all over the globe as upper-atmospheric currents carried the debris from a land impact. The climate would rapidly cool. Violent blizzards would rage where the weather had been benign and constant just a few days before. Eventually, the dust would settle to the surface, the sky would clear, and temperatures would moderate again—but not until monumental changes had been wrought on Earthly life.

It is fun to speculate about how a cosmic event might affect life on Earth via psychological, rather than brute-force physical, means. We might never be able to prove that this sort of indirect catastrophe did (or did not!) ever cause a great change in the course of history, but it is not hard to imagine how a comet could scare primitive people into doing things they would not do if no comet appeared. Consider the following example.

A comet, as brilliant as the quarter Moon and having a tail visible even in full daylight, passes near the Earth. Stone-Age men, women, and children see the phenomenon and become terrified that their land will soon be smashed by a force from above. So they decide they must move far away to a place more safe. Other Stone-Age families get together with them and form a caravan, unprecedented in history, migrating to a far-away land of fertile plains and rivers. There they build a city, the first of its kind, realizing that human beings can survive more easily when they work together as compared with hunting and gathering individually.

Several thousand years ago, the historians tell us, our ancestors did undergo this fundamental change in their mode of thinking. Cities arose. Some might argue that this would have happened sooner or later regardless of whether any cosmic display occurred. Nevertheless, it could have been triggered by an eclipse of the Sun, a supernova— or a comet apparition.

CHARACTERISTICS OF MEMORABLE COMETS

Most comets never get bright enough to be seen by an ordinary person without the aid of binoculars or a telescope. The ones we remember are the ones we can plainly see with naked eyes.

Chapter 2 describes how the brightness of a celestial object can be measured. This is easy to do with stars and planets (which are essentially point sources of light). In the case of a comet, brightness is harder to determine because various comets have tails of different lengths and heads of different sizes. A small, dim comet with a short tail might actually seem brighter than a large, diffuse comet with a long tail.

Other factors that affect our perception of how spectacular the object is include the shape, the altitude of the comet above the horizon and the orientation of its tail with respect to the horizon. Because of these variables, some of the most famous comets have been dimmer, smaller, or have had shorter tails than less well-known comets. Usually, a long-tailed comet is more likely to be remembered (for a given brilliance) than a short-tailed one. This is true even though the smaller comet actually looks brighter because its light is more concentrated in the field of view.

Before the development of precision astronomical measuring devices, there were many comets that were mentioned in historical writings. We have to be subjective when we compare such apparitions to present-day or recent comets such as Halley's Comet in 1910 or Ikeya-Seki in 1965. We can get some idea, although vague in some cases, of their characteristics from the exhaustive descriptions given.

The length of a comet tail was often expressed in early literature in terms of "feet," which has been assumed by some to correspond to degrees of arc in the sky. This might not be what was meant by the term. A "foot" might have meant the apparent length of a foot ruler laid on the ground at one's feet (A of Fig. 3-1), which would mean an arc span of about 10 degrees. More likely, the ancients would have held out their hands at arm's length and measured the distance from hand to hand. A "foot" would then mean about 25 degrees of arc (B of Fig. 3-1). There is obviously a large uncertainty factor in these various interpretations: 10 to 1 or, more probably, 25 to 1! So we do not really know how long the tails of long-ago comets really were as they appeared in the sky. In more recent times, tail spans in excess of 60 degrees have been observed.

The relative brightness of comets, in ancient times before measuring apparatus was available, could only be compared to other celestial objects such as Venus or the Moon in a certain phase. Some comets have evidently reached apparent brilliances approaching that of the full Moon (although such apparitions are rare).

We might expect that estimates of comet size and brightness by ancient astronomers has been overly exaggerated in some cases. This is because history gets distorted with the passage of time. The phenomenon of exaggeration is well-known to psychologists. Perhaps you are familiar with the experiment wherein a tale is told to a person at the end of a long line, the story is relayed along, and at the other end of the line, the last person produces a completely different version of the original tale.

Comet heads generally appear smaller in size

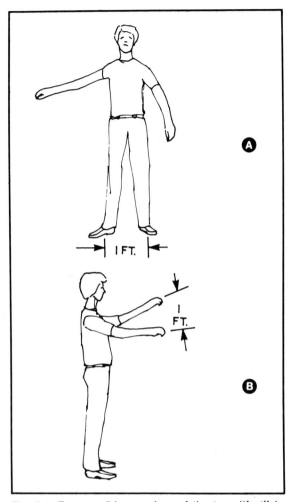

Fig. 3-1. Two possible meanings of the term "foot" in reference to angular measure. At A, the apparent span of a foot ruler laid at one's feet. At B, the apparent span of a foot ruler held at arm's length.

than the full Moon. The more concentrated light of a comet could mislead even the most astute observer into judging the brilliance of a comet head to exceed that of the full Moon. In reality the comet head was perhaps much less brighter than that.

In A.D. February 1106, an especially bright comet was seen in Europe. Although its tail was relatively short, according to the records, if we assume that a "foot" is 1 degree of arc (it was "a foot and a half in length"), its magnitude has been estimated at -10. That is about the same as the quarter Moon. It was clearly visible in daylight.

Another bright comet, estimated to have a magnitude of -5 (comparable to a crescent Moon), was seen in 1402. Another especially brilliant comet, DeChesaux, was seen in 1744, having a magnitude of about -5. Perhaps one of the brightest of all comets was 1882 II, which might have appeared nearly as bright as the full Moon. Its tail was rather short, estimated at less than one degree of arc.

Comets generally reach maximum brilliance at or near perihelion. Depending on the way in which the Earth, comet and Sun are arranged at this time, a potentially famous comet might not appear spectacular at all. Some of the brightest comets, known as the Sungrazers, have been observed during total solar eclipses and found to be exceedingly bright. Nevertheless, the brilliant photosphere would overwhelm the comet.

NAMING OF COMETS

Comets were originally not named. Thus we still hear of such natural events as the Great Comet of 1861, the tail of which the Earth passed through. As astronomical telescopes became larger (they could gather more light), comets were discovered so often that it became impractical to let them go unnamed. Adjectives weren't enough. Comets were christened according to their initial discoverers.

Naming a comet after its discoverer can cause two problems. What if two or more different people discover the same comet at about the same time? This problem is solved by allowing a comet to bear as many as three names. The names are separated by hyphens. Thus we have had comets such as Ikeya-Seki and Arend-Roland. It has been decided that more than three names would be ridiculous and cumbersome. Therefore, we do not hear of comets such as Jones-Smith-James-Olson-Walters-Peterson-Garcia-Welch!

The other problem is more difficult to solve from a scientific rather than egotistical or legal standpoint. One astronomer might discover two, three or even more comets in his or her career. This has led to a system of naming comets according to the year in which they are first observed. The comets then are renamed according to their perihelion dates once their orbits have been calculated.

Suppose that the first comet seen in 1999 is found on January 30, and the next is seen on March 1. Then the January comet receives the name 1999a, and the March comet is called 1999b. Subsequent comets would be called 1999c, 1999d, and so on.

What happens if it is later discovered that 1999a and 1999b were actually the same comet, discovered independently by observers in different countries? Then the whole series of comets for 1999 will have to be renamed. This is done once the perihelion dates are known for each comet. The letters are replaced with Roman numerals. Comet 1999a/b would thus be known as 1999 I, assuming perihelion was to occur in 1999. If perihelion was not to take place until 2000, then comet 1999a/b would be renamed comet 2000 I. Comet 1999c might be changed to 2000 I or 2000 II, or, if an intermediate comet was discovered, perhaps 2000 III. Comet 1999d might become 1999 I or 2000 III, or whatever.

Eventually all comets get unique names. New comets (those not seen before) are also named after their discoverers. Lay people use the discoverers' names more often than the numerical designators when referring to comets; we think of comet Ikeya-Seki, not 1965 VIII, for example. Astronomers, too, use the discoverers' names much of the time, unless there is a need for more specific information.

The numerical system of naming comets has gotten rid of most of the confusion surrounding past comet apparitions. The Great Comet of 1861 is now known among astronomers as 1861 II; it was the second known comet to have passed perihelion in 1861. This comet, discovered by John Tebbutt of New South Wales, is sometimes called Tebbutt's comet. This kind of ambiguity of names, all representing the same comet apparition, can cause confusion. A perfect naming system might never be invented. To this day the strange objects befuddle us.

THE MOST FAMOUS COMETS

Throughout recorded history, but especially since

just before the birth of Christ, astronomers began to notice comets and record their characteristics. Every few years, a fairly bright comet would come along. Perhaps once in a lifetime a truly spectacular apparition would be seen. One or more ancient astronomers would be likely to witness at least one great comet. Some of the most noteworthy comet apparitions are described in the following paragraphs.

Comet of 240 B.C.

According to historical records, the earliest known apparition of the object we now call Halley's Comet was in 240 B.C. There is some evidence that suggests that the comet was brighter (in absolute terms) than it is now.

Comet of 146 B.C.

We must deal with the problem of interpretation from ancient languages to the modern as well as possible exaggerations of how brilliant comets were in the past. After a great war, an unusually spectacular comet was seen during 146 B.C. The comet was described by some observers as comparable to the Sun in size (presumably the coma of the comet was that large). This would imply that the comet came very near the Earth. The comet of 146 B.C. might have been the original of the "Sungrazer group" (see comets 1843 I and Ikeya-Seki).

Julius Caesar's Comet

After the murder of Julius Caesar in Rome in 44 B.C., a comet of unusual brilliance appeared in the sky. The comet was at times so bright that it could be viewed in full daylight. Some philosophers and religious people of the time believed that the comet was the soul of Caesar being transported into the heavens by the gods.

Comet of A.D. 1066

The apparition of Halley's Comet in A.D. 1066 preceded the invasion of England by the Norman French. It seems that many Anglo-Saxons took it to be a premonition of disaster. The French might have been spurred on by it. This demoralizing effect on the defenders, and the possible boost in confidence it gave to the aggressors, probably did not affect the outcome of the war. We cannot say for certain. If it did, we might owe many of our English words today to an astronomical object!

Comet of A.D. 1077

A comet or other brilliant object, visible in daylight near the Sun, was seen by the English in A.D. 1077. Some astronomers have suggested that it was not a comet, but Venus near inferior conjunction—when that planet appears at its brightest. Today we know that some comets are occasionally seen very close to the Sun, even in broad daylight, with appropriate viewing apparatus. And Venus, after all, goes through inferior conjunction often enough that we must wonder why A.D. 1077 would have been such a special year.

Comet of A.D. 1106

A spectacular comet, approaching the brilliance of the quarter Moon, is described in A.D. 1106 writings as appearing first in the morning, and later in the evening. The object was bright enough near perihelion that it could be seen in broad daylight.

Comet of A.D. 1222

Historical writings from the Far East describe a comet visible in full daylight. This was evidently an apparition of Halley's Comet, according to extrapolations backward in time. Of course, we can never be certain of this because brilliant comets have been seen just before or after apparitions of Comet Halley.

Great Comet of A.D. 1264

The great comet of A.D. 1264 was visible with the naked eye even after sunrise. It is remarkable for its long tail, which spanned more than half of the sky at one time, estimated to have been at least 90, and perhaps as much as 100 degrees in length as seen from Earth. European and Far Eastern records give similar descriptions of the comet. We can be fairly sure that it was indeed an especially spectacular sight.

Comet of A.D. 1402

A brilliant comet was visible in daylight for several days in A.D. 1402 as it passed through an orbit favorable for viewing from Earth. This comet appeared in March and, near the Arctic Circle, is said to have been seen continuously, day and night, for at least several days.

Comet of A.D. 1472

According to Chinese records, a comet became so bright during A.D. 1472 that it was visible all day long. This is apparently not the result of its having an especially bright absolute magnitude, but rather that it passed close to Earth.

Comet of 1577

Discovered by Tycho Brahe, the comet of 1577 became visible in full daylight.

Discovery of the Telescope. Just after the seventeenth century began, astronomers began to use small telescopes to view the heavens. This made it possible to see comets that would have been invisible previously. When we look at the historical records, we must make a distinction between those apparitions seen before and after about A.D. 1600. It is possible, in fact likely, that astronomers would react much more strongly to a rather unspectacular comet after the invention of the telescope as compared with the period prior to the seventeenth century A.D.

Comet Kirch, 1680

The object, first discovered by the English astronomer G. Kirch in 1680, eventually acquired a visible dust tail about 90 degrees in length, spanning half the sky as seen from our planet. This comet might have been a member of the Sungrazer group, perhaps the same object as Ikeya-Seki in 1965.

Great Comet of 1744

The great comet of 1744 passed very close to the Earth, as evidenced by its fanned-out dust tail. At one time, there were 11 separate, identifiable tails; six of them were easily visible with the naked eye. Figure 3-2 is a drawing of what this comet might have looked like at the time it was the most spectacular. This comet is sometimes known as DeChesaux's comet (named after its discoverer). With the aid of a telescope, the head of the comet could be seen in full daylight for a short time.

Comet 1843 I

Even when it was near the Sun in the sky, the unusually bright comet of 1843 could be seen in broad daylight. At its peak, it appeared about as bright as the quarter Moon. The tail spanned more than 70 degrees of the sky at its greatest apparent length. This comet is believed to be one of several comets in a common orbit having extremely small perihelion distances. Comet Ikeya-Seki might be another member of the Sungrazer group.

Comet Biela, 1846 and 1852

First seen in 1772, Comet Biela reappeared in 1805. Many astronomers suspected that it would reappear at intervals of a little less than seven years; they were correct. Biela's Comet showed a period of about six years and nine months in its elliptical journeys around the Sun.

In 1846, Biela's comet was seen through a telescope as having a double head—as though the nucleus had broken in two. This had not been observed on previous apparitions. A few days later, the division had become unmistakable. There was not one comet, but two—side-by-side with parallel tails. The two objects moved together, and one comet was somewhat smaller than the other. Figure 3-3 is a drawing of how the double comet looked according to descriptions. Could the two objects be separate comets? Astronomers did not think so because of their proximity (which was maintained).

In the 1852 apparition, the comet was again seen to have two parts. They moved together, and—just as in 1846—one was smaller and dimmer than the other. The comet was not seen after 1852, and astronomers concluded that it must have

Fig. 3-2. The Great Comet of 1744 as it might have appeared with six prominent tails.

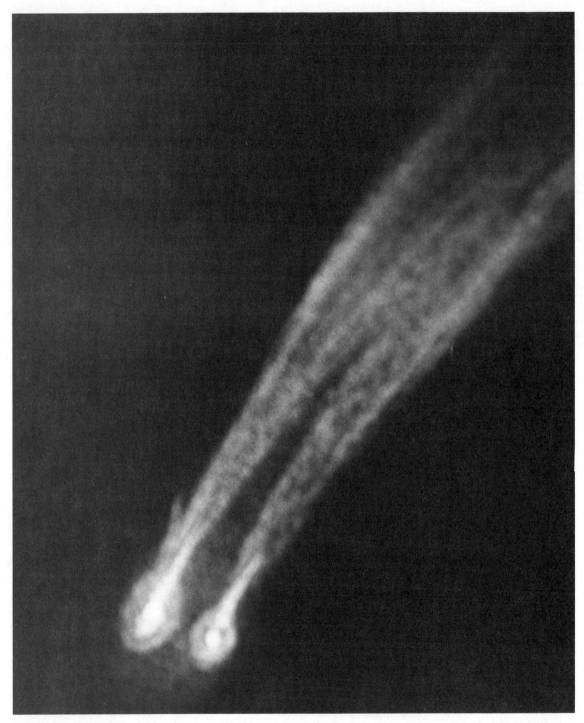

Fig. 3-3. Hanging in the sky like a strange celestial being, Comet Biela split in two during the 1846 apparition. This drawing probably shows how it looked as the breakup became noticeable with the unaided eye.

disintegrated totally. In 1872, at the time the comet would supposedly have been seen, an awesome meteor shower appeared. Most astronomers think that this shower, never previously seen at that time of year, was caused by the remnants of Biela's comet.

Great Comet of 1861

The Great Comet of 1861, discovered by Tebbutt and designated 1861 II, came extremely close to Earth (as did the comet of 1744). This gave the tail a strikingly fanned-out appearance as the comet passed between the Sun and the Earth. For a time, the Earth was actually within the tail of the comet. This provoked fears among some people that adverse effects such as climate changes or meteorite bombardment might occur. Nothing of this kind happened.

Comet 1862 III

Comet 1862 III, discovered by L. Swift, was interesting because of its Sunward-pointing spike. Dust tails generally point only away from the Sun as seen from the Earth. The exact reason for the appearance of the Sunward spike is not known with certainty, but it is believed to be an effect of perspective. There is good evidence for this theory. See the discussion and the photographs of comet Arend-Roland that follow.

Great Comet of 1882

The second comet to reach perihelion in 1882, the Great Comet of 1882 ranks among the brightest comets ever seen. It was about as bright as the quarter Moon, and some of its tail could be seen— even without the aid of a telescope or field glasses—while the Sun was above the horizon.

Comet 1910 I

Not to be confused with Halley's Comet, Comet 1910 I reached perihelion in January of the year 1910. It was a fairly dim comet in terms of absolute magnitude, but its small perihelion distance and its close passage to the Earth made it visible for a time in daylight.

Halley's Comet, 1910

The 1910 apparition of Halley's Comet was better suited to viewing from the Southern Hemisphere than from the Northern Hemisphere, but provided an interesting sight all over the world. Halley's Comet is not the most spectacular ever seen, but it is famous because it was the first comet shown to have appeared on previous occasions. In the 1910 apparition, Halley's Comet appeared with a straight dust tail and a large, prominent coma. More is said about Halley's Comet later in this chapter.

Comet 1914 V

Comet 1914 V was discovered during the last weeks of 1913, and it remained visible for several months during 1914. Superstitious people believed the comet foretold of the coming of the Great War (World War I). It was a naked-eye object during the nighttime hours.

Comet Skjellerup, 1927

Comet Skjellerup was best observed in the Southern Hemisphere at night. As it passed perihelion, however, it was visible in full daylight everywhere except north of the Arctic Circle. (In that part of the world in December, when the comet was near perihelion, the Sun is always below the horizon.) The tail could be seen with the naked eye near the Sun's disk.

Comet 1947 XII

There were no exceptionally bright comets found for 20 years after the apparition of Skjellerup's Comet. During 1947, however, a promising object became visible to astronomers in the Southern Hemisphere. It was a vivid, orange naked-eye object. The peculiar color of this comet was evidently the result of abundant sodium in the head. Sodium vapor, when excited by radiant energy, produces the familiar glow of the candle flame.

Comet Arend-Roland, 1957

Near the end of 1956, a comet was discovered by two Belgian astronomers, and it was named after them. The comet passed perihelion around the middle of 1957, becoming easily visible with the unaided eye and developing a long brilliant tail. The interesting feature of this comet is the Sunward-pointing "antitail" or "spike." Figures 3-4A, 3-4B, 3-4C, and 3-4D show the comet Arend-Roland on four separate occasions in April and May, 1957, as seen through the 48-inch telescope at Mount Palomar, California.

It is generally thought that Sunward-pointing tails are simply a result of the angle at which a comet is viewed from the Earth, along with an unusual amount of gas or dust in the plane of the comet's orbit. Figure 3-5 shows how this strange visual illusion might be produced as a comet passes in between the Earth and the Sun. We would suspect that the Sunward "spike" would be visible only for a rather short time if the perspective theory were right. And this is just what did happen. After a few days, the "antitail" was gone because the comet had moved enough to eliminate the effect.

Comet Mrkos, 1957

A Czechoslovakian astronomer discovered a comet during 1957 that turned out to be an interesting object. This comet, called Mrkos after its first observer, holds the distinction of being discovered without the aid of a telescope or field glasses. Under exceptional viewing conditions, Mrkos found it with the unaided eye. Other naked-eye observers found it, too, before it was noticed by people equipped with telescopes!

The comet Mrkos developed a highly intricate tail, billowing as though it were a plume of smoke or long tresses of wispy hair blowing in a breeze. This effect is shown in Figs. 3-6A, 3-6B, 3-6C, and 3-6D that were taken through the 48-inch telescope at Mount Palomar. The billowing dust tail became spectacular, as shown in Fig. 3-6A, around August 17. The straight, vivid rays of the gas tails are clearly visible (Fig. 3-6B) along with the complicated dust tail, which even seems to have an eddy. It seems that the solar wind must really represent moving particles, just as atmospheric wind consists of air molecules in motion. One would almost think that the dust tail of Mrkos' comet looks like smoke! Different perspective views are shown in Figs. 3-6C and 3-6D. The dust tail is long and tenuous; the gas tail appears brighter, straighter, and shorter.

Comet Ikeya-Seki, 1965

The eighth comet to reach perihelion in 1965, Ikeya-Seki (named after its two Japanese discoverers) was unusually brilliant before sunrise. Near perihelion, it reached an apparent visual magnitude brighter than that of the quarter Moon, and it was visible in daylight. Even the tail could be seen through a telescope near the Sun during daylight hours. Just after sunrise, the comet was a spectacular red color.

Because of the extremely small perihelion distance of Ikeya-Seki, and the similar paths of a few other comets observed in the past, some astronomers think Ikeya-Seki, and these other comets, might be fragments of an original massive object that was broken up by the tremendous heat of the solar corona. These smaller comets would gradually move away from each other along the original orbit (Fig. 3-7). This hypothetical family of comets is called the Sungrazer group. From descriptions of past comets, it appears that Ikeya-Seki might lie along the same orbit as comets that appeared in 146 B.C. and A.D. 1843. It is even possible that one or both of those comets actually were Ikeya-Seki at past apparitions.

Comet Bennett, 1970

Just before New Year's day, 1970, J. Bennett found a small comet south of the celestial equator. At about the same time, D. Seargent found the same comet. By March 1970, it was about 50 million miles from the Sun—well within the orbit of Venus—and its tail could be seen easily with unaided eyes. The gas and dust tails were plainly identifiable. The astronomer D. Seargent observed and described the appearance of this comet at length.*

---

* David A. Seargent, *Comets: Vagabonds of Space* (Doubleday & Co., Inc. (1982), pp. 150-155.

Fig. 3-4A. Comet Arend-Roland, as seen on April 26, 1957, had a striking sunward "spike" (courtesy of Palomar Observatory, California Institute of Technology).

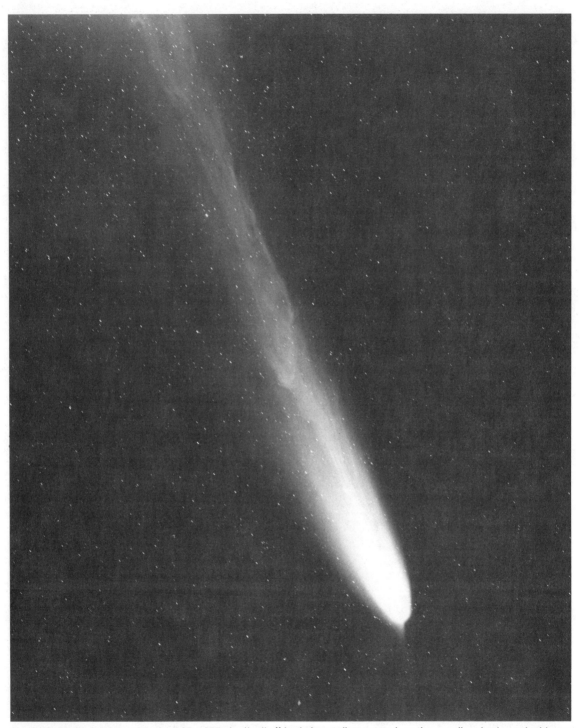

Fig. 3-4B. By the following day, April 27, 1957, the "spike" had almost disappeared, as the peculiar viewing coincidence passed (courtesy of Palomar Observatory, California Institute of Technology).

Fig. 3-4C. Comet Arend-Roland as seen on April 29, 1957. The plane of the ejected gas, which was probably responsible for the "spike," is well outside the line of sight. Almost all evidence of its existence has disappeared. The part of the gas tail opposite the Sun is, however, plainly visible in this photograph as a set of bright, straight lines (courtesy of Palomar Observatory, California Institute of Technology).

Fig. 3-4D. Arend-Roland on May 1, 1957. The dust tail has spread out. All evidence of the strange sunward "spike" is gone. There are several tail components visible (courtesy of Palomar Observatory, California Institute of Technology.

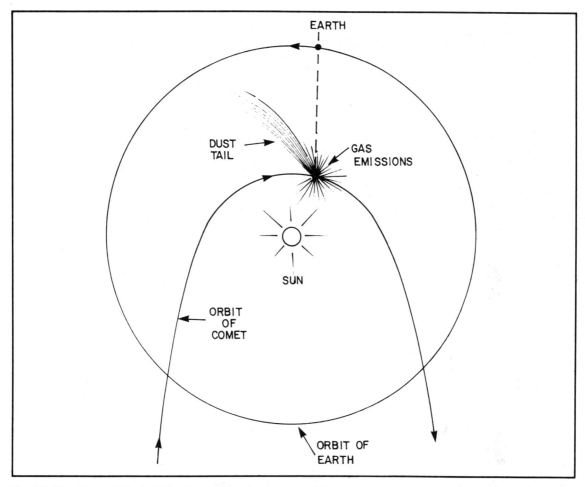

Fig. 3-5. The Sunward "spike" that is occasionally seen protruding from the head of a comet could be caused by an unusual amount of ejected gas in the plane of the comet's orbit. The "spike" would be most prominent as the Earth passed through the plane of the comet's orbit.

The nucleus and much of the tail appeared yellow, characteristic of energized sodium vapor. Close observation of the nucleus showed gas and dust spirals suggestive of rotation. By the end of April, the tail had grown to its maximum length of about 25 degrees of arc.

Comet Kohoutek, 1973-74

The comet called Kohoutek is notorious because of its failure to live up to expectations. It was discovered early in 1973 by L. Kohoutek of the Hamburg Observatory. Like many comets, this one was found by comparing photographs during a routine survey of the heavens. The perihelion was to occur at the end of 1973 and the beginning of 1974, and the nucleus was to pass within 10 million miles of the Sun. Kohoutek thus promised—or seemed to promise—to become a brilliant comet, perhaps more spectacular than any other previous twentieth-century comet.

As the comet approached the Sun from well outside the orbit of Mars, the brightness increased rapidly. The comet became a well-known news item to the general public. I recall taking an astronomy course at that time, and the instructor told us that

Fig. 3-6A. Comet Mrkos, August 17, 1957. The main feature of this comet was its wispy, complicated tail (courtesy of Palomar Observatory, California Institute of Technology).

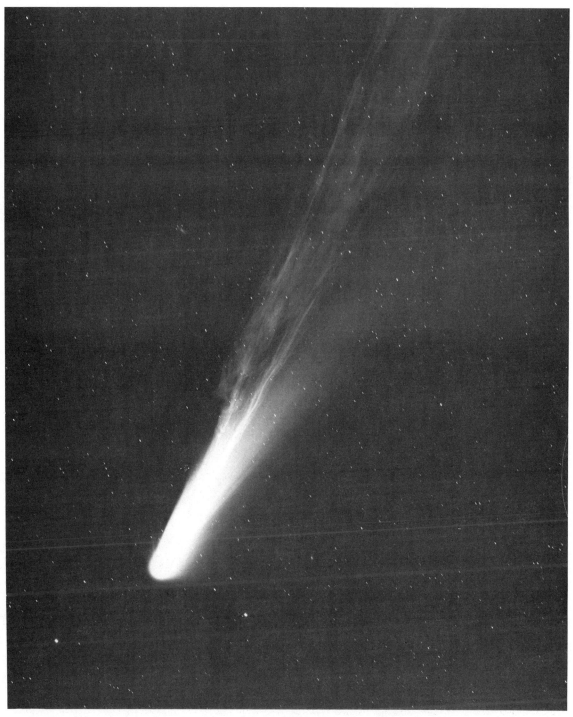

Fig. 3-6B. Comet Mrkos, August 22, 1957. The gas tail is visible at the right, and the dust tail at left. There is evidence of eddy-like activity within the dust tail (courtesy of Palomar Observatory, California Institute of Technology).

Fig. 3-6C. By August 26, 1957, the gas tail of Comet Mrkos had fallen almost exactly into line with the dust tail (courtesy of Palomar Observatory, California Institute of Technology).

Fig. 3-6D. Comet Mrkos on August 27, 1957. The dust tail had thinned out and become essentially straight as seen from our planet (courtesy of Palomar Observatory, California Institute of Technology).

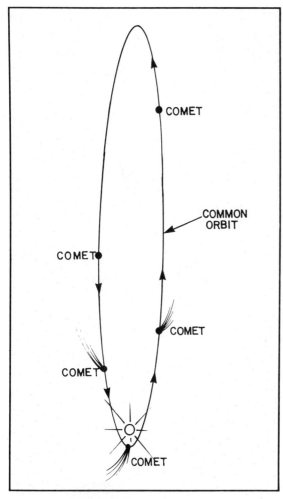

Fig. 3-7. Several different comets might follow a single orbit around the Sun.

Kohoutek would probably be a dominating feature of the sky as it passed perihelion. But it was not to be.

The rapid increase in brightness did not continue. Instead, as the nucleus plunged past the orbits of the Earth, Venus, and finally Mercury, the comet seemed to peter out. The comet was easily visible with a telescope or field glasses (Figs. 3-8A, 3-8B, 3-8C, and 3-8D), but was difficult to see with the naked eye.

Why did this happen? Why didn't Comet Kohoutek become as bright as had been anticipated—rivaling the quarter Moon in brilliance even well after sunset? The reason is a mystery, but perhaps it had something to do with the structure of the nucleus. There might have been a core consisting mainly of rock, and almost no icy matter, covered with a relatively thin layer containing almost entirely ice (A of Fig. 3-9).

If that were true, then the ices in the outer layer would have burned off long before the comet reached perihelion, leaving only the inner core containing only a little ice. Because it is the icy material, and not the rock, that is mainly responsible for the glow of the coma and the expanse of the tail, this theory gives a plausible explanation for the failure of Comet Kohoutek to achieve the glory that so many people (some astronomers as well as laymen) believed it would.

Another, similar, theory is that the concentration of ice diminished gradually from the surface to the center (B of Fig. 3-9). It could also be that the nucleus was of lower density than those of most other comets and, therefore, it burned away much more rapidly than normal.

Comet West, 1976

The disappointing surprise of Comet Kohoutek during 1973 and 1974 taught us that comets do not always behave according to predictions. In 1976, the lesson was repeated, but in the opposite, more spectacular, fashion.

Comet West was found by photographic means. It was at first assumed that the comet would not achieve great brilliance because this often happens with comets discovered on photographs. The comet remained in view long enough for the orbit to be determined. The nucleus was to reach perihelion at the end of February, 1976, at a distance of less than 20 million miles. Comet West had the potential to be spectacular.

After the experience with Kohoutek, astronomers were reluctant to make a prediction of exceptional brilliance for Comet West. Near perihelion, with the aid of field glasses or a telescope, the nucleus and part of the tail were visible in daylight.

Early in March the comet became observable

Fig. 3-8A. Comet Kohoutek, as seen through the 48-inch Schmidt telescope on October 31, 1973. The tail was just beginning to develop (courtesy of Palomar Observatory, California Institute of Technology).

Fig. 3-8B. Comet Kohoutek on November 24, 1973. The tail was plainly visible through a telescope, but the comet was still undetectable with the naked eye (courtesy of Palomar Observatory, California Institute of Technology).

Fig. 3-8C. By December 1, 1973, the tail of Comet Kohoutek had split in two. The comet was, however, still a telescopic object (courtesy of Palomar Observatory, California Institute of Technology).

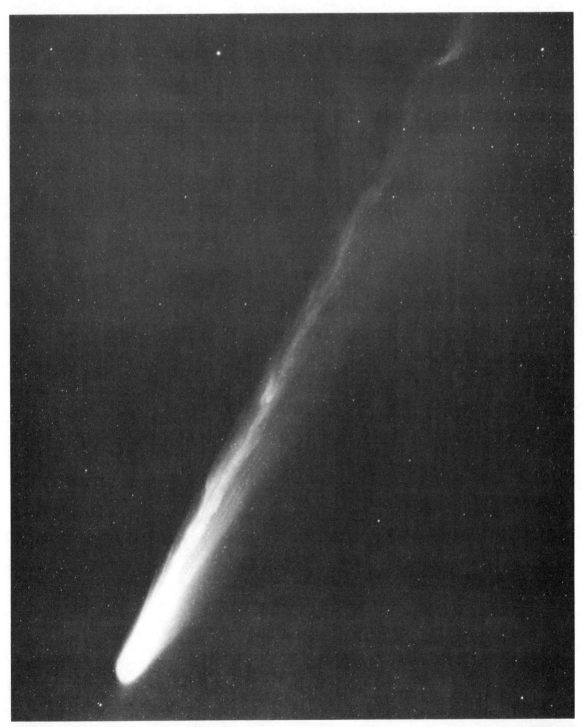

Fig. 3-8D. Comet Kohutek on January 12, 1974. A fairly long tail finally developed, and the object was dimly visible with the naked eye under ideal viewing conditions (courtesy of Palomar Observatory, California Institute of Technology).

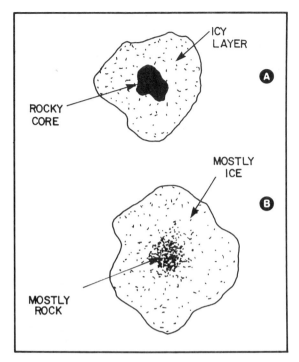

Fig. 3-9. Here are two possible explanations for the failure of comet Kohoutek to achieve expected brillance. The nucleus might have had a solid, rocky core surrounded by almost pure ice (A), or it might have had a greater concentration of rock near the center than the surface (B).

in the predawn sky. The tail was quite long and the coma was brilliant. Like many other comets, it exhibited a yellowish color characteristic of sodium. Then the nucleus split in two. Four days later it had broken into quarters. One of the nuclei moved rapidly away from the others, apparently blown around by its own ejected gas.

It is quite possible—even likely—that you have never seen a comet even if you were alive during the apparitions of such comets as Ikeya-Seki, Bennett, and West. These were all naked-eye objects, quite spectacular at their maximum brilliance, and yet, although I was 12, 16, and 22 at those times, I never saw any of the three comets. That might be because I lived in a town large enough so that the lights obscured diffuse astronomical objects. No doubt, many people miss morning comets unless they are early rising semiinsomniacs.

Geography can also interfere with the viewing of comets. Those who live in valleys surrounded by large hills or mountains, for example, cannot see low enough in the sky to notice a comet in darkness or twilight. Parts of the world are overcast much of the time, and that makes any kind of astronomical viewing inconvenient. Persistant haze or air pollution are also detrimental to comet watching.

In order to see a comet, it is usually necessary to first find out about it—to know it exists! That often requires some effort because newspaper headlines don't often say things like "GREAT COMET VISIBLE THROUGHOUT NORTHERN HEMISPHERE." In a few cases, comets do make prominent news; Halley's surely will. You must know when and where to look. Will it be an evening or morning object? You also must have good viewing conditions. A large pair of binoculars or perhaps a small telescope at low magnification are helpful.

HISTORY OF HALLEY'S COMET

Scientists did not always know that comets are often periodic (they travel in closed orbits around the Sun, reappearing at regular intervals). As recently as the late seventeenth century, it was not generally known that comet apparitions, occurring only days apart on opposite sides of the Sun, might represent the same object. Isaac Newton—famous for his gravitational theories—when told by a colleague that such a double apparition was a single comet, refused to believe it for five years. This in spite of his own contribution to the theory of celestial orbits!

Edmond Halley, born in 1656, was destined to prove conclusively that a certain comet had reappeared again and again throughout history at intervals of about 76 years. His proof came according to the scientific method: A theory must be demonstrated true by showing that it accurately predicts physical happenings.

Halley (who evidently preferred the pronunciation "Hawley" and not "Hailey," as most of us say it now) had the good fortune of a large allowance, provided by his father, during his research years. He traveled to the Southern Hemisphere to complete his star catalog—which he finished at the tender age of 22—and became known as a famous astronomer. Halley helped Newton publish his famous *Principia* and other books; he was secretary of the Royal Astronomical Society in England for a time. By all indications, he was an extremely energetic man, and he used all of his abilities to their utmost. Finally, Halley was made head of the geometry department at Oxford University.

Halley's research with comets was characteristically methodical and thorough. His fascination with the fleeting, ethereal objects became almost a preoccupation. His position at Oxford served mainly to provide him with the free time needed to pursue the mysteries of comets.

Halley noticed from astronomical records that there was a resemblance among several comet apparitions in the past. In particular, he noticed that the comets of 1531, 1607, and 1682 seemed remarkably similar. Might they all have been recurring apparitions of a single object, moving in a greatly elongated elliptical orbit around the Sun? If this were true—and he had yet to prove it—the comet must have an aphelion, or maximum distance from the Sun, of about 37 astronomical units. That was well outside the confines of the known Solar System; Uranus, Neptune and Pluto had not yet been discovered.

Halley continued to dig through the historical records to see whether or not there were reports of comets appearing at fairly regular intervals before 1531. He found the evidence promising. Today, we know of numerous comet apparitions (Table 3-1) that were probably caused by returns to perihelion of the comet we have named after Halley.

If his theory were correct, Halley reasoned, then the comet should return in the year 1758. It turned out that the estimate was a little early because nobody knew that Uranus and Neptune, both having extensive gravitational fields, existed. These two planets perturbed the orbit of the comet, and it was not first observed until Christmas of 1758. The object went through perihelion in the late winter of 1759, and began its long journey back into the cold depths of the Solar System during the spring. Halley did not live to see his prediction verified; he died 17 years earlier.

Edmond Halley was, in some sense, lucky to have made his discovery. Today we know that two or more comets can follow roughly the same orbit. If the comets of 1531, 1607, and 1682 had been such objects, like the sungrazer group, Halley would have been misled. Nevertheless, the rapidly increasing extent of knowledge would eventually have resulted in the discovery of the periodic nature of this prominent comet. Perhaps the truth about this comet would not have been known until 1910; we might then be calling it Herschel's Comet!

By the time the nineteenth century waned, astronomical calculations had been improved to the extent that the perihelion date for return of Halley's Comet could be forecast to within one day. The search for the comet began in 1908, but the first positive observation was not made until September

Table 3-1.

Apparition Number, Relative to 240-239 B.C.	Year of Apparition
1	240-239 B.C.
2	163 B.C.
3	86 B.C.
4	11 B.C.
5	65-66
6	141
7	218
8	295
9	373-374
10	451
11	530
12	607
13	684
14	760
15	836-837
16	912
17	989
18	1066
19	1145
20	1222
21	1301
22	1378-1379
23	1456
24	1531
25	1607
26	1682
27	1759
28	1835-1836
29	1910
30	1985-1986

1909. (Later it was found that the comet had actually appeared in photographs taken a month earlier.) The perihelion date was calculated as April 19, 1910 by two English astronomers, Crommelin and Crowell.

The 1910 apparition of Halley's Comet was more spectacular after perihelion than before, and more easily seen in the Southern Hemisphere than in the Northern Hemisphere. Photographs taken of the comet showed a rather straight and uniform tail. The most spectacular displays were observed in early May. See pages 200, 202, and 204.

Halley's Comet in 1910 was less spectacular in the Northern Hemisphere (where most of the world's population lives) than many comets that had been seen in the previous hundred years, but it was of special interest to astronomers attempting to accurately pinpoint the perihelion time. Perihelion occurred on April 19, 1910, exactly as predicted half a year before.

ORBIT OF HALLEY'S COMET

When Edmond Halley observed the great comet of 1682, he attempted to calculate its orbit and found that the object evidently followed a parabolic path. Actually, we now realize, the comet moves in an elongated ellipse. When we see just a small part of a very eccentric ellipse, the curve is nearly identical with that of a parabola (Fig. 3-10). It was only after noting the strangely uniform spacing of great comet apparitions over the years that Halley began to realize the true orbit of the comet. Its aphelion is more distant than Neptune, and the perihelion is inside the orbit of Venus. Figure 3-11 illustrates the orbit of Halley's Comet in relation to the planetary orbits in the Solar System.

Halley's Comet orbits outside the plane of the Solar System. The plane in which the comet moves is tilted at about 18 degrees relative to the ecliptic, and the comet revolves in the opposite direction from all the planets. During most of its orbit, the comet is south of the ecliptic. Only near perihelion, for a short time, does it swing north of the ecliptic. Because of this, a collision between Halley's Comet and any of the outer planets—Jupiter, Saturn, Uranus, Neptune, or Pluto—is impossible as long as the orbit remains constant. An encounter with Mars is also impossible under current conditions. Only a small change in the orbit of the comet would be necessary, however, for a catastrophe to be possible with Venus or Earth.

As Halley's Comet falls in toward the Sun past the orbit of Mars, it passes through the ecliptic plane and moves into the Solar System's "Northern Hemisphere." It remains there through perihelion—until the comet reaches approximately the orbit of Venus—whereupon it once again enters the Solar System's "Southern Hemisphere." There it stays until the next return 76 years later.

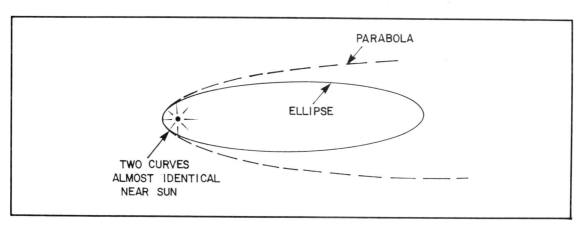

Fig. 3-10. When a comet's orbit is calculated according to points close to the Sun, it is difficult to tell whether it is an ellipse or a parabola.

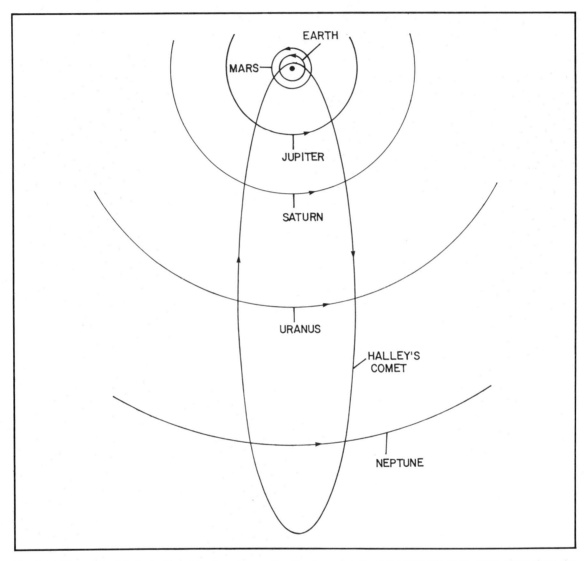

Fig. 3-11. The orbit of Halley's Comet is viewed from directly above (north of) the Solar System. Most of the time, the comet is south of the ecliptic plane Only near perihelion does it swing north of the ecliptic.

Only if an ecliptic crossing—either the one before perihelion or the one after—were to occur at precisely 93 million miles from the Sun would it be possible for a collision to take place with the Earth. With Halley's Comet, neither ecliptic crossing occurs at the same distance from the Sun as the Earth's orbit. A slight perturbation, however—perhaps caused by Jupiter, Mars, or even our own planet—could change that. Then, on some return of the comet in the distant future, a catastrophe might happen.

WHAT WILL WE SEE IN 1985-86?

Those who expected an awesome sight in 1910 were amply rewarded as Halley's Comet approached and passed perihelion. During that apparition, the Earth actually passed through the tail of the comet. This caused concern among the uniformed who feared horrible effects from such an astronomical coincidence. Nothing unusual happened on our little planet that could be attributed to the comet, unless we are willing to stretch our imaginations enough to blame a few heart attacks, strokes, or suicides on certain individuals' hysteria about the mysterious celestial apparition.

With the 1985-86 return, the Earth will be farther from the comet and will not come close to its tail. The closest the comet will come to us is about 39 million miles. The prophecies of the doomsayers, few as they may be in the enlightened third quarter of the twentieth century, cannot be borne out. Halley's Comet will not strike our planet!

For viewers in the Northern Hemisphere, Halley's Comet will probably not become a naked-eye object until New Year's Day, 1986. At that time, the coma should be dimly visible near the celestial equator between the constellations of Pegasus and Capricorn. With the aid of binoculars

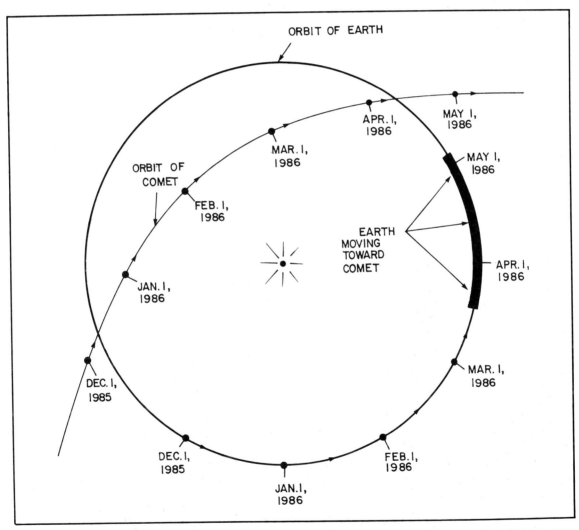

Fig. 3-12. Approximate positions of Halley's Comet relative to the Earth between December 1, 1985 and May 1, 1986. From about mid-March to early May, 1986, the Earth will be constantly moving closer to the comet. This is when we can expect to get the best view.

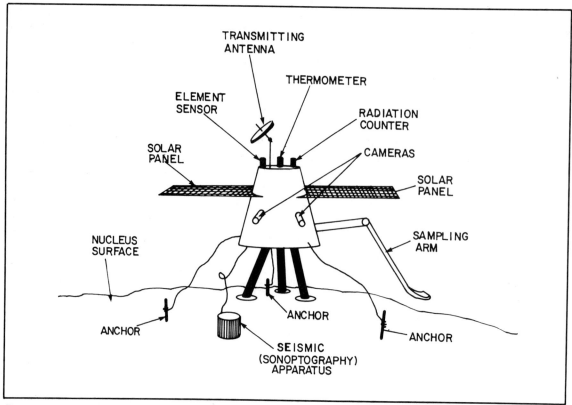

Fig. 3-13. A hypothetical, unmanned probe for landing on the surface of a comet nucleus.

or a small telescope, people should be able to pick up the comet (near the constellation Pegasus) in late November or early December 1985. At that time, the comet will look like a fuzzy, slightly elongated star—similar to a distant galaxy or nebula—but a little brighter.

By the middle of January 1986, the tail should be clearly visible, when using binoculars, as the comet hangs low in the western evening sky. As it approaches perihelion, the comet will be lost in the evening twilight, and its tail will actually seem to get shorter as it points more and more directly away from us. The comet will pass around the Sun on the far side, and not on its near side as was the case in 1910. Therefore, we will not see the comet's tail when it is at its longest.

The best view of Halley's Comet for Northern-Hemispheric observers will be in the morning sky after perihelion. At that time, the earth will be racing more or less toward Halley's Comet (Fig. 3-12), and we will get a fairly good view of the tail because of the favorable vantage angle. Unfortunately, the comet will be moving southward after its brief visit to the northern side of the ecliptic plane, and it will be wintertime in the Northern Hemisphere of the Earth. Therefore, the tail will not be ideally situated with respect to the horizon.

Those who wish to get a good view of Halley's Comet should make an effort to get well away from populated areas where lights and haze tend to muddle the atmosphere. A good, low view of the eastern horizon will be necessary to catch the tail in its maximum splendor during the middle and latter part of March. Those who can afford it might consider booking a trip to some Southern-Hemisphere country to view the comet under better astronomical circumstances. The tail will point almost directly upward in the morning sky during

March and April, as seen from such places as Australia, New Zealand, and southern South America.

Halley's Comet will make its closest approach to the Earth—a distance of 39 million miles—during April. The tail will be getting rapidly shorter by then, but the proximity of the comet will render it fairly spectacular for much of that month. As spring advances, the comet will fade from naked-eye view and become, once again, a telescopic object as it embarks on its long trip back out into the cold, dark, far reaches of our Solar System. It will not return to perihelion again until July, 2061.

VISITING COMETS

Plans have been made by various governments for sending unmanned probes to Halley's Comet in 1985-86. This would give us a closeup look at the nucleus, and might (at least partially) resolve the question of what really makes up a comet.

France, Japan, and the Soviet Union have seriously entertained the idea of sending probes to approach, and perhaps land on, the nucleus of Halley's Comet. So far there have been two main ideas:

☐ Come close to the nucleus and photograph it, from a distance of a few hundred miles.
☐ Crash-land on the surface, in a manner similar to the first lunar probes, taking photographs at intervals until the very end.

The first plan would allow investigation of the behavior of the comet's nucleus for a period of months (from long before perihelion until long afterward). The second plan would only let us see the comet for a short time, but we would get a truly close-up view.

A third possible comet-probe mission would involve a soft landing. Figure 3-14 illustrates a hypothetical unmanned probe as it might conduct experiments from the surface of a comet nucleus. The unmanned landing mission, while much more difficult than the other two plans, would provide the most detail about the composition of the nucleus. Direct samples could be taken, and perhaps they could even be brought back to the Earth for laboratory study.

The interior of the nucleus could be analyzed by seismic methods; shock waves could be sent through the material and their speeds and echo characteristics evaluated. Seismic analysis has provided us with a good idea of the interior structure of our own planet. When we talk about landing on a comet's nucleus, we are assuming that the Whipple model is more or less correct. A spacecraft couldn't touch down on a diffuse swarm of pebbles.

A fourth possibility exists for the long-term future: put an astronaut on a comet. Right now this is out of the question. Even if we had sent people to Mars (which we haven't, yet), it would be unreasonable to dispatch someone to visit an object the composition of which is largely unknown. Perhaps we will have journeyed to a comet by the early or middle part of the twenty-first century.

Unfortunately, by the time you read this, it will probably be too late for any new probes to be sent to Halley's Comet for the 1985-86 return. What has been sent will be all. It takes time for a spacecraft to traverse millions of miles at the speeds we can muster today. Even if we could accelerate a probe to extreme speeds, reaching a comet in a few days or even hours, the problem of accuracy—and the possibility of a complete miss—would be greatly increased as compared with slower speeds.

You don't just aim a space probe at a celestial object as though you were about to shoot a pop gun at it. The paths of the comet and the probe must be carefully matched, bit by bit, and course corrections made at intervals as necessary. The interaction of gravitational fields in three dimensions presents a problem far more complex even than that faced by a football quarterback as he attempts to complete a long pass to a wide receiver while both men are running and a strong wind is blowing. As any quarterback will tell you, such a task is hard enough!

We can leave the orbital calculations to the computers. What is a comet made of? How do they originate? Why do they have such elongated orbits?

A close look at a comet's nucleus might help us better answer these questions.

A TRIP TO A COMET'S HEAD

What would it be like to fly a spacecraft to the nucleus of a comet, land there, and ride out the perihelion? If Whipple's dirty-snowball model is correct, we can imagine such a hypothetical adventure.

We would have to choose a comet of moderate or large size with a perihelion distance great enough to avoid excessive temperatures. (We would get broiled on a Sungrazer.) The comet would have to possess very little spin; a rapidly rotating comet might throw us off into space (besides making observations more difficult). Halley's Comet might be a good choice for a manned expedition. Because it has returned many times in the past without disintegrating, it would be unlikely to fall apart under us. Let's imagine ourselves as the occupants of a manned vessel on its way to rendezvous with Halley's Comet in the year 2061.

We approach the slowly tumbling, grayish-white, irregular mass of the nucleus before a tail has begun to develop. Even as far away from the Sun as the orbit of Mars, there is already some evidence of evaporation taking place in the volatile ices. We pick a spot on the surface that is to our liking, and shoot out several harpoonlike anchors to attach ourselves to the object. Then we touch down. Three long cables must be wrapped all the way around the nucleus to ensure that we will not be ceremoniously dumped off into space in the glaring heat of perihelion (unless we are forced to make an emergency exit). We know that the nucleus will probably shrink as the Sun blows some of the comet's matter into space. The cables will have to be periodically tightened.

We pass the orbit of the Earth. The Sun slowly rises and sets over the eerie, nearby horizon. There is practically no gravity because the comet is not only small, but of low density. As the spacewalkers complete the tethering, they know this well enough. A good, all-out high jump could toss a human being out of the gravitational field of the comet forever.

On the comet surface, things are starting to happen. During the comet's "day," while the Sun is up, white and yellow streamers poke up briefly and settle back, then race toward the horizon like dry-ice mist being blown along by a fan. During the comet's "night," the sky is filled with a diffuse glow, obliterating all but the brightest stars. This glow is the newborn tail, seen along its whole length, that will eventually grow to millions of miles. The sunrise and sunset remind us of early morning and late afternoon of a foggy winter day on our home planet.

The comet draws nearer to the Sun, and the days grow brighter but progressively more hazy. The porous surface erupts, furiously now, in a blinding blizzard of gases and particles. During the comet's "morning" and "afternoon," the winds rage in a thin atmosphere made up of methane, ammonia, water vapor and other gases in trace amounts. At comet "noon," the wind temporarily abates as the Sun crosses the zenith. During the "night," the sky is ablaze with turbulent, glowing vapors that seem to sweep upward from the horizon in all directions and converge on the point in the sky exactly opposite the Sun. Earth observers radio us to say that the tail has become long and quite spectacular.

The nucleus has begun to shrink somewhat, and the cables girdling it have gone slack near our vessel. We pull them in enough to make them reasonably tight again. The harpoonlike anchors have come undone from the vaporizing icy surface. They are drawn in and fired again. An astronaut is sent out to check the condition of our vessel. Conditions are too severe for him to stay out very long; the blowing gas and dust makes it nearly impossible to maneuver, and the heat of the Sun is becoming too intense for spacewalks.

Perihelion is reached, and the storm on the nucleus rises to its peak of fury. During the "daytime," it is impossible to see anything at all except a white and yellow blur. During the "night," the display of the coma and tail is overwhelming. The sky is so bright at "night" that a book can be easily read by its light.

Now the comet is moving away from the Sun, we have passed perihelion! We will have to wait about three months before we can lift off and rendezvous with the mother ship somewhere past the orbit of the Earth. The time goes by slowly. We are anxious to get back home and evaluate our findings. The storm on the nucleus gradually clears up, and the "night" sky fades with each revolution until we can once again see the brighter stars during the hours of darkness. Finally, we recognize the disk of the Sun through the swirling vapors during the comet's "day." We go outside to take many photographs and a look at what has happened to the nucleus during the passage.

The familiar landmarks of the preperihelion surface are mostly gone. The small "hill" near our vessel has completely disappeared. The tethering cables are slack in places, proving that the nucleus has shrunk. We figure its mass has decreased by about 1 percent, but we will not know exactly un-

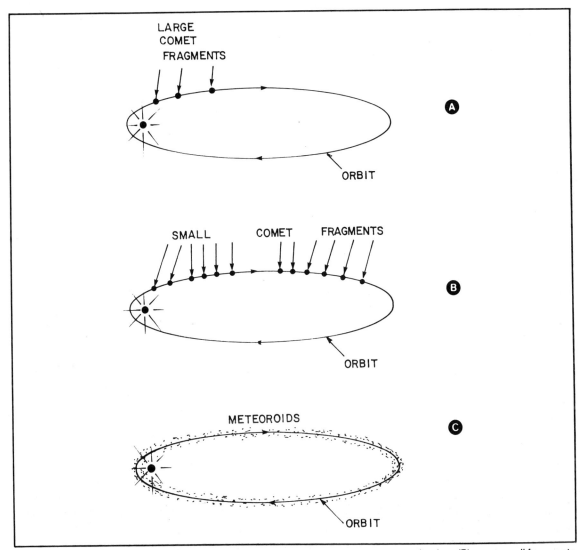

Fig. 3-14. The gradual breakup of a comet: (A) three large fragments, fairly close to each other; (B) many small fragments, spread out over a large portion of the common orbit; (C) the final degenerative stage where a swarm of meteoroids is distributed almost uniformly around the whole orbit.

til more precise observations are made and evaluated.

Suddenly, almost instantly, there is a flash behind us that rivals the Sun in brilliance. Not 200 yards away a tremendous geyser has erupted! Is it a comet flare-up? We hastily make our way back to the ship and dispatch a message to our Earth-based observers, asking them if the comet has gotten brighter. The reply arrives within the hour. Yes, the coma has brightened from magnitude 7 to magnitude 6—more than a twofold increase. The tail has grown from 3 degrees to 7 degrees of arc in the sky. We were standing, it would seem, right next to the origin of the flare-up!

The adventure of the past few months must draw to a close a little earlier than previously planned, says the commanding officer of our landing vessel. Our engineer has been monitoring the seismic activity within the comet, in conjunction with sounding tests, and for the first time she reports a massive disturbance. We cannot remain on the surface of this comet; the whole nucleus might be splitting in two! We have about two hours to carry out the final preparations for leaving the comet. The tethering cables are pulled apart at the splice points. (Someone had the foresight to design them so that, in the remote event of a catastrophe on the nucleus of the comet, the cables would not be a problem).

The ship is disengaged from the cables. A second, even more monstrous, eruption has begun. If this comet is going to break up, says our engineer, it will happen soon. We'll get a better—and safer—view from a distance.

We lift off from the surface and move out several kilometers, coming to rest on the solar side of the nucleus. We will be following the comet and watching what happens to it. There are now four geysers spewing matter into space. As the comet rotates, the wisps of gas and dust are twisted into bizarre pinwheels before the solar wind disperses them.

A piece of the nucleus breaks away and floats out ahead of it. Then another chunk breaks loose and slowly moves Sunward. Now we can see, firsthand, how a comet gradually breaks up into constituents, all of which follow essentially the same orbit as the original. The rotation of the main nucleus imparts more or less velocity to the fragments that are thrown off. Gradually, over thousands of years and dozens of orbits, the pieces spread out (Fig. 3-15). Ultimately, perhaps a million years or more after the comet's first plunge Sunward, nothing is left but a swarm of meteoroids following an elongated elliptical orbit. This is what gives us on Earth our annual displays of meteor showers, such as the Leonids, Orionids, and Perseids.

Finally, the prediction of our engineer is realized. A fantastic eruption is taking place all around the perimeter of the mass of ice and rock. Earthbound observers are reporting a significant increase in the brightness of the coma and the length of the tail. This is unusual for a comet nearly one astronomical unit from the Sun. Slowly, the nucleus separates into two pieces of nearly equal size. We can actually watch them as they move apart. After several hours, the separation is obvious even with the naked eye. Halley's Comet has followed the course that all comets eventually must follow. A turning point has come. When Halley's Comet reappears in 76 more years, our grandchildren will see not one comet but two.

So our hypothetical journey to Halley's Comet ends. Of course, we can't really say that this comet actually will split in 2061, but it is quite probable that—sometime in the next ten, hundred, or million centuries—it will. Perhaps our descendents will then have an adventure.

Chapter 4

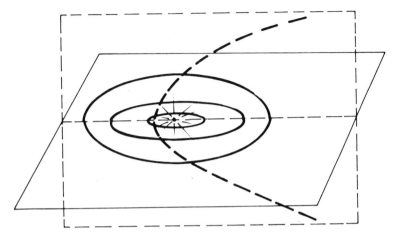

In Search of Comets

SEVERAL NEW COMETS ARE FOUND EVERY year. For every truly new comet that is discovered, several old ones are located and re-tracked. Sometimes we find that a previously observed, supposedly one-time comet is actually periodic (recurring). In a sense, this is like discovering a new comet. Chapter 3 describes how one astronomer became famous by finding that a particular comet was periodic, orbiting the Sun once every 76 years. Edmond Halley, after whom we have named the comet that will appear in our skies in 1985-86, successfully predicted that the comet would return over and over again.

At first thought, it would seem that the search for comets is strictly up to the professional. Certainly, amateur astronomers or casual observers cannot sit in a backyard and scan the heavens with binoculars or a small telescope, hoping to find a new comet—can they? Yes they can. Even the most expert astronomers, using the most advanced telescopes and photographic apparatus, miss comets that amateurs notice. This is primarily because comet hunting is a time-consuming pursuit, and the major observatories simply do not have the time to devote to it.

Many amateur astronomers set aside several hours a week solely to look for new comets. Their equipment is not all that sophisticated or expensive. Small telescopes or field glasses have yielded many new comet discoveries. In 1957, the Czechoslovakian astronomer Mrkos discovered a comet with unaided eyes!

COMET-SEARCHING DEVICES

Of course, no one is likely to make momentous astronomical discoveries just by going to the top of a hill, waiting for a half hour until the eyes adjust to the darkness, and then staring off into the cosmos. The pupil of the eye dilates to about a quarter of an inch in darkness, a small pair of field glasses provides an aperture of an inch and a half, and a small telescope has an opening of 2 to 4 inches in diameter. These viewing devices thus collect tens or hundreds of times as much light as unaided eyes. Figure 4-1 shows the relative light-gathering power of various aperture diameters com-

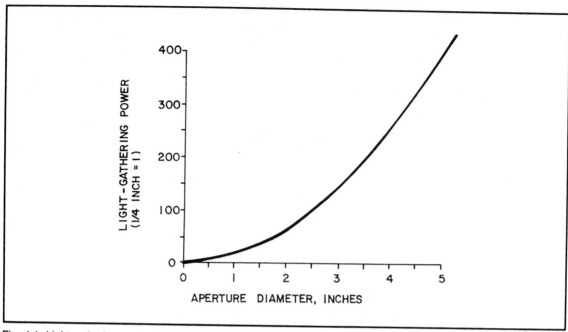

Fig. 4-1. Light-gathering power of a telescope relative to the pupil of an eye dilated to one-quarter inch.

pared with an unaided eye having a pupil a quarter of an inch across.

Another important specification of a telescope or pair of binoculars is the magnification factor. Magnification is very important when it is necessary to resolve features of a distant planet. In many cases, however, magnification (often referred to somewhat inaccurately as "power") is not so crucial to good observation. Sometimes magnification might even hinder the ability of the astronomer to resolve detail. Comet hunting is one of those pursuits that does not require a great deal of magnification.

Magnification is defined in terms of apparent linear size. This is based on the idea that a telescope or binoculars makes things look a certain number of times "closer." If two stars appear to be 1 minute of arc (1/60 of a degree) apart as seen by the naked eye, then a magnification factor of 3 would make them seem 3 minutes (1/20 of a degree) apart; a magnification factor of 6 would make them seem 6 minutes (1/10 of a degree) apart; and so on.

Binoculars generally have magnification factors in the range of 5 to 10. Small telescopes can provide magnification factors of several hundred. The largest telescopes can, in theory, magnify distant objects many thousands of times (although atmospheric turbulence limits the resolution that can be obtained).

Magnification is defined in linear terms. For objects such as the Moon, the Sun, a planet, or a comet, doubling the magnification results in a four-fold increase in the apparent area we see. Tripling the magnification causes an increase of nine times in the apparent area. A dim, hazy object becomes harder and harder to see, with a given telescope, as the magnification is increased because the available light is distributed over a large area with high "power" as compared with low "power."

If we use too much magnification, we might miss a diffuse celestial object altogether due to its light being so spread out that we cannot distinguish it from the blackness of space. This problem can occur with comets that have already developed a coma and perhaps a short tail. For this reason, amateur comet hunters generally use binoculars or telescopes with fairly low magnification. But sometimes relatively high "power" is needed. If a comet is first seen when it is still far away from the Sun—say near the Asteroid Belt—it might look like an ordinary star under low magnification. Only by using high "power" would the embryonic coma and tail become recognizable.

The magnification of a telescope or field glasses depends on two things: the focal length of the objective lens or mirror, and the focal length of the eyepiece. Focal lengths are commonly expressed in millimeters (mm). The magnification factor is simply the ratio of the focal lengths of the objective and the eyepiece (Fig. 4-2).

For example, a pair of binoculars with objective lenses of 56mm focal length and eyepieces of 7mm focal length would have a magnification factor of 56/7 (or 8). A telescope with a 1000mm objective lens and a 10mm eyepiece will have a magnification factor of 1000/10 (or 100). Most telescopes have an eyepiece socket that can accommodate eyepieces of various focal lengths to provide different degrees of magnification.

When we want to observe diffuse objects, such as nebulae, galaxies, or comets, light-gathering power is of more importance than brute-force magnification. The more light we can get from a dim, distant object, the more detail we will be able to see because human eyes are actually not very sensitive to light. Therefore, a large-diameter telescope is much better for comet hunting than a small-diameter one. An aperture of at least 2 inches is desirable. Most binoculars are smaller than this, but some military-surplus warehouses have larger ones. Binoculars have certain advantages (discussed shortly) over telescopes despite the generally smaller size of binoculars.

HOW TO COMET HUNT

If you are interested in searching for comets, you

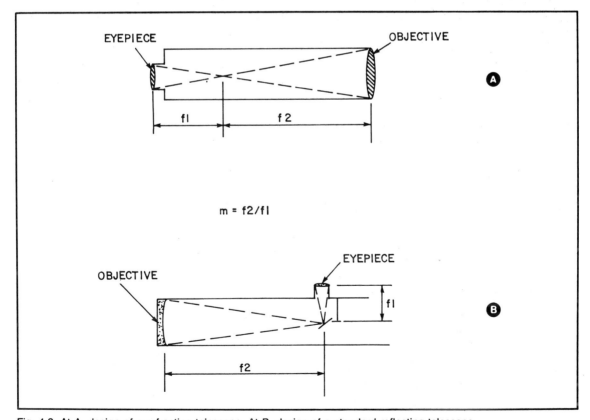

Fig. 4-2. At A, design of a refracting telescope. At B, design of a standard reflecting telescope.

are not alone. Hundreds of amateur astronomers engage in this hobby on an almost nightly basis. And, it should be repeated, amateurs make significant discoveries of previously unknown comets. What should you do if you are interested in comet hunting?

Move to the Country!

Except for perhaps the viewing of the Sun and Moon, cities wreak total havoc on all astronomical observations. The scattering of light by the molecules of air alone is bad enough. The additional pollution in the air over a large city, as compared with county air, worsens the problem. Color filters and polarizing devices are of little help. For best observation of dim and diffuse celestial objects (like comets), you must get well away from cities.

This does not really mean, of course, that you have to move to the country. It is nice if you have some friends there that you can visit while you observe the heavens. Those who already live in the country, or have a second home there, can consider themselves fortunate in this respect!

Altitude

Most major observatories, such as Mount Wilson, Palomar, and Kitt Peak, are located atop mountains. This is no accident. The less air there is between you and the heavens the better your view will be because atmospheric dust, pollution, and turbulence scatter the light from distant celestial objects. You have no doubt seen how a star, such as Sirius or Betelgeuse, seems to "twinkle," while a planet, such as Venus, Mars, or Jupiter, does not.

The "twinkling" effect is the result of refraction in the atmosphere of our own planet (Fig. 4-3). This limits the resolving power of any telescope, no matter how powerful, because it blots out fine detail. On a mountain top, there is less air to interfere with astronomical observation as compared with a sea-level place. If you are very serious about comet hunting, you should try to conduct your pursuits from a location that is at a high altitude.

Cloudiness

Any serious amateur astronomer will have a frustration level that is proportional to the cloudi-

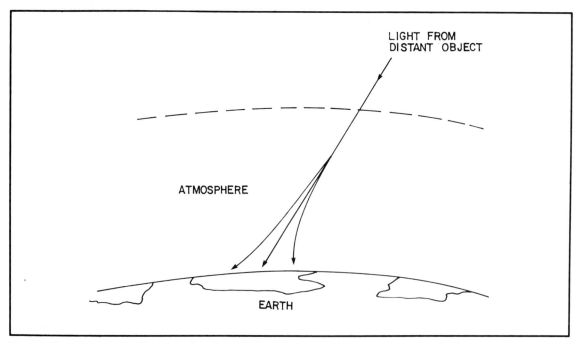

Fig. 4-3. The atmosphere refracts the light from distant celestial objects, limiting the obtainable resolution.

ness of the weather. Some places, such as the desert Southwest in the United States, have clear skies much of the time. Many other places are less fortunate from the point of view of the astronomer. A few places are overcast quite a lot of the time, and these areas are obviously the least well-suited for gazing at the heavens.

Humidity

To some extent, the amount of water vapor in the air adds to the scattering of light from the stars. High humidity is often accompanied by a certain amount of haze. A dry climate provides better viewing conditions ("all other things being equal") than a wet climate. High humidity can also contribute to fogging of lenses in telescopes and binoculars.

Temperature

You aren't going to enjoy your comet hunting very much if you have to sit still for hours when it is 40 degrees below zero. Temperature is, to some extent, an important factor for the astronomer. It is rarely too hot for comfort during the night in any part of the world, but cold can be an issue.

Excessive cold is bothersome and it can cause problems with telescopes and binoculars. Fogging of lenses occurs very easily when the temperature drops below about 50 degrees Fahrenheit. At temperatures below freezing, frost can form on the lenses. Condensation or frost is most likely to form when an instrument is brought indoors after being subjected to low temperatures, but it can also accumulate outdoors if you breathe anywhere near the lenses.

When dew or frost forms inside a telescope or a pair of binoculars, getting rid of the moisture can be difficult. Most telescopes can be opened up and cleaned. With binoculars this is often not possible. For the benefit of the observing apparatus as well as the person, a moderate temperature is best for any kind of astronomical work.

Letting Your Eyes Adjust

It takes about a half hour for normal eyes to adjust to the darkness of a moonless night, allowing the maximum amount of light to enter the pupil and the greatest retinal sensitivity. Even a short glance at a bright light, such as a street lamp or oncoming automobile, will desensitize the eyes for several minutes once they have gotten fully adjusted to the dark. No serious observation should be done when your eyes are not at their best.

Amateur astronomers often find it necessary to consult written or printed material while observing the heavens. For example, you might want to check the position of a certain fuzzy object, which might be a comet, to be sure it is not a nebula. You need a light to see the star charts, but even the light of a lantern, reflected from a piece of paper held in your hand, can temporarily desensitize your eyes. The use of lights should be kept to a minimum, and lights should be as dim as possible.

Sometimes it is absolutely necessary to use a light. The eye is less sensitive to red light than to other colors. Therefore, a red color filter (or a red bulb) will reduce the extent of eye desensitization. A piece of clear red cellophane, or even a red cloth, will do nicely when placed over the lens of a flashlight or lantern.

Binoculars or Telescope?

Binoculars offer one major advantage for the aspiring comet discoverer. It is much less fatiguing to keep both eyes open, and to see the skies with both eyes, than to constantly squint through the monocular eyepiece of a conventional telescope. A pair of binoculars also offers a wider field of view than a telescope because binoculars have objective lenses with shorter focal length in relation to their diameter (Fig. 4-4).

The main disadvantage of binoculars is their relative lack of magnification. While brute-force magnification of several hundred times is not usually needed for comet hunting, small binoculars are sometimes not powerful enough. If you want to be the first to discover a comet, you will almost certainly have to find it while it is still quite far away from the Sun. Through a low-magnification pair of binoculars, such an object could easily be mistaken

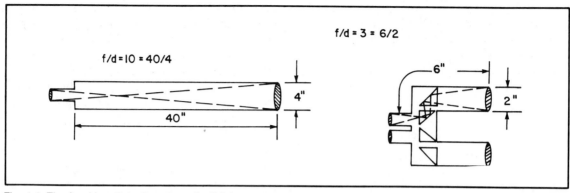

Fig. 4-4. The focal-length-to-diameter ratio (f/d) is generally larger for a telescope (A) than for a pair of binoculars (B). The values shown here are typical but they vary from one instrument to another.

for a star; the coma and tail will not be very well developed. If a telescope is used, however, sufficient magnification can be obtained to resolve suspicious-looking objects. All you would have to do is exchange eyepieces to increase the magnification when necessary.

Some comet hunters attach binoculars to a telescope, aligning them so that objects appear centered in both fields of view (Fig. 4-5). With this kind of setup, the sky can be scanned with a minimum of eye fatigue and a maximum field of view using the binoculars. When a potential comet is spotted, a quick look through the telescope will usually resolve it.

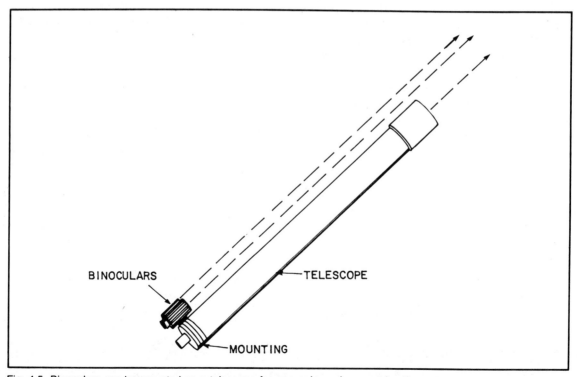

Fig. 4-5. Binoculars can be mounted on a telescope for convenience in comet hunting.

For those who can locate one, a binocular telescope will work very well for comet searching. These devices are similar to the large binoculars sometimes found in parks, at beaches, or overlooking cities. Binocular telescopes offer the advantages of ordinary binoculars as well as conventional telescopes.

Refractor or Reflector

From the standpoint of light-gathering power, a reflecting telescope is no better or worse than a refractor of the same size. A reflector and a refractor, both having the same diameter and focal length, will have equal resolving power (assuming that the objective lens or mirror are not flawed). Some amateur astronomers prefer reflecting telescopes; others would rather use a refracting telescope.

For the systematic sky searching that is necessary in comet hunting, a refractor is somewhat easier to use than a reflector. This is because, when using a refractor, your eyes are looking in the same direction as the telescope is pointed (unless a right-angle prism is used at the eyepiece). With most reflecting telescopes, you are looking away at a right angle from the object under observation. It is therefore harder for some people to be sure they are not missing certain portions of the sky.

There is one type of reflecting telescope that allows you to look in the same direction that the instrument is pointed. This device employs a convex mirror and a hole for the eyepiece at the center of the objective mirror (Fig. 4-6). This type of telescope is known as a Cassegrain or Cassegrainian reflector. For amateur astronomy, Cassegrainian reflectors are excellent. They can be made physically short with large objective diameter, providing a wide field at low magnification while still allowing higher magnification if desired. Such an arrangement is ideal for comet hunting.

A reflecting telescope does enjoy one important advantage over the refractor for comet hunting: a mirror, silvered on the outside (called a first-surface mirror), reflects light at the same angle regardless of color and regardless of how short the focal length might be. This is not true of a glass lens; it refracts light at different angles at different wavelengths. This phenomenon, known as dispersion, gets more and more pronounced as the focal length is made progressively shorter (Fig. 4-7). Violet light is refracted more than green light, which in turn is refracted more than red light.

Lens manufacturers use special coatings and lens designs in an effort to minimize dispersion, but the effect can never be totally eliminated. Therefore, you will not often find refracting telescopes with short focal lengths that allow wide-field observation. The dispersion problem would blur every star, planet, or other celestial object—making them look like little color spectra. With a reflecting telescope that has a first-surface mirror, the focal length can be made as short as the state

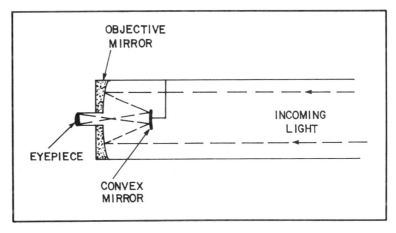

Fig. 4-6. Design of a Cassegrainian reflector.

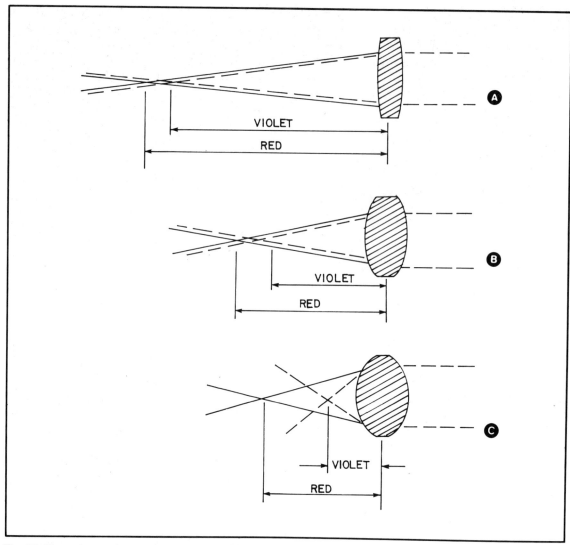

Fig. 4-7. Dispersion, or the tendency for a lens to refract different colors of light at different angles, increases as the focal length decreases. At A, a long focal length causes little dispersion. At B and C, shorter focal lengths result in greater dispersion.

of the mirror-grinding art will allow. And there will be no dispersion problem.

At large observatories, reflectors enjoy a decided advantage over refractors. Huge lenses are difficult to support at the top end of a long tube. In practice, a lens of more than about 40 inches diameter is impractical because a lens can be supported only around the edges. Cross braces cause stellar objects to appear x shaped.

Another problem with large refractors is caused by dispersion: the focal length has to be very long if a large-diameter lens is used or color problems will occur. With a reflector, the objective mirror can be supported from behind—without interfering with the light entering the instrument—and the focal length-to-diameter ratio can be much smaller because a first-surface mirror is free from dispersion effects. Figure 4-8A shows a refracting

Fig. 4-8A. A large refracting telescope (courtesy of U.S. Naval Observatory).

Fig. 4-8B. A large reflector (courtesy of U.S. Naval Observatory).

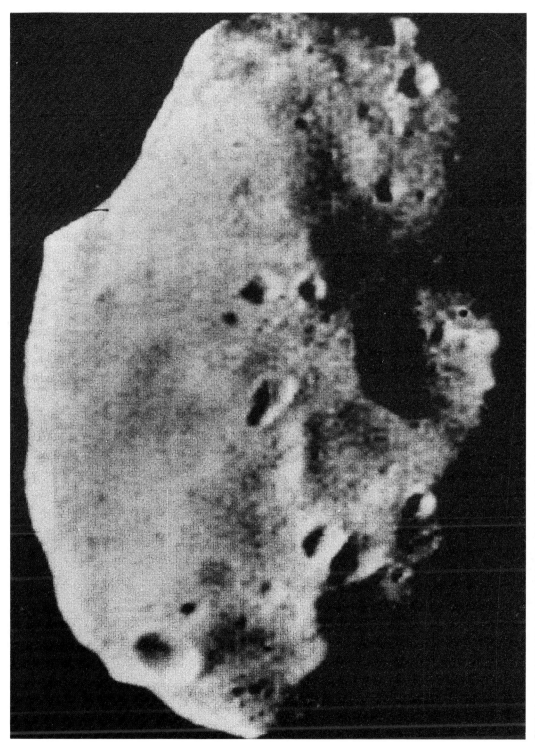

Phobos, one of the two moons of Mars—Deimos is the other—was discovered in 1877 by American astronomer Asaph Hall. The satellites are very small, irregularly shaped ellipsoids. Mars is named after the Roman god of war. Phobos (fear) and Deimos (terror) are named for the mythical horses that pull the war god's chariot. (NASA photograph.)

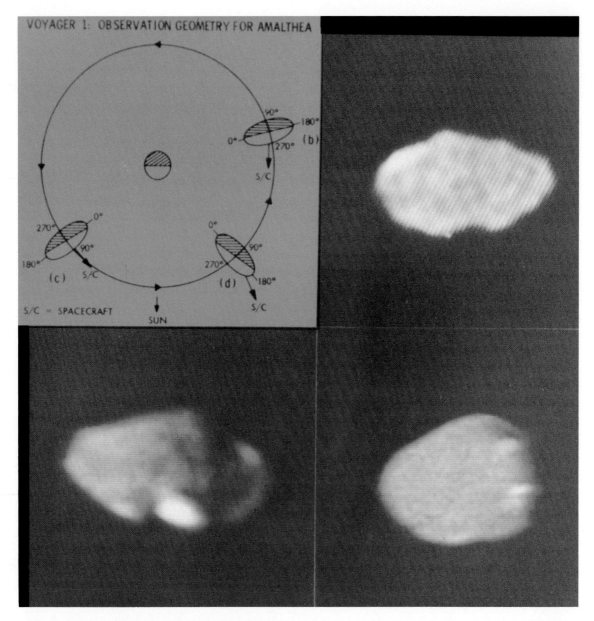

Three views of Amalthea, a tiny moon of Jupiter, as seen from the Voyager 1 space probe. Amalthea is actually an asteroid, probably captured by the gravitation of Jupiter after straying from the main "belt" of asteroids. The tiny satellite is irregular in shape, and appears to vary in color depending on the angle of viewing. Jupiter probably has thousands of smaller, still undiscovered asteroid moons. (NASA photograph.)

telescope with a diameter of 26 inches and Fig. 4-8B shows a reflector with a diameter of 61 inches. The difference is plainly evident.

If you are serious about comet hunting, ultimately you will have to decide whether the advantages of the refractor outweigh those of the reflector for your purposes or whether you prefer the reflector.

Type of Mounting

Whenever a magnification factor of more than about 10 is used, it becomes necessary to mount a telescope or binoculars on a firm base. If you want a vivid demonstration of the reason for this, try looking at the Moon, a planet, or a star with a telescope of high magnification without using a tripod or pedestal mounting.

There are two basic schemes used by astronomers in mounting telescopes. The first, and simpler, type is known as the az-el (azimuth-elevation) system. With an az-el mount, the telescope can be moved horizontally and vertically (A of Fig. 4-9). This up-down, left-right arrangement is easy to "get the feel of" because it is a natural way to perceive coordinates in space. The other mounting scheme is known as the equatorial system. This is essentially an az-el mount that is tilted so that the azimuth rotation is in the plane of the celestial equator (B of Fig. 4-9). Using this kind of arrangement for casual observation is a little more difficult than using the az-el system because the equatorial mount is actually aligned for the poles—90 degrees north and south latitude—and not very many people are going to conduct their observation from those places.

For serious astrophotographic work, an equatorial mount is a necessity. The equatorial system is also desirable for prolonged observation of a single object. This is simply because, as the Earth rotates, it is easier to follow a celestial object with an equatorial mount as compared with an az-el system. Some equatorial mounting apparatus is equipped with a clock drive that keeps the telescope pointed at the same object in the sky for minutes or even hours. Once the mounting is properly aligned, so that the azimuth rotation precisely follows the celestial equator, correction is made by turning the telescope westward at the rate of one revolution every 23 hours and 56 minutes (the sidereal rotation period of the Earth). At the major observatories, this method is used to obtain long exposures of distant nebulae, galaxies, and other faint objects that cannot be seen while looking directly through the eyepiece.

Despite all of the advantages of the equatorial mount for observation of known objects, this arrangement is not very good for comet hunting. An equatorial-mounted telescope is great for looking at one part of the sky for a long time, but when you embark on a comet search you have no idea where (or even if) you might come across your quarry. An equatorial mount tends to be rather confusing. Perhaps this is because our intuition or sense of space perception is better suited to the az-el mount. It is easier to go up and down and across—relative to the local horizon—than to move according to some distant latitude.

The Search

It takes, on the average, hundreds of hours for a well-equipped amateur to find a new comet. There have been coincidences in which neophyte comet hunters made a discovery after just a few hours or even on their first night. Even for them, over a period of time, the law of averages catches up.

The brightest comets usually are seen near the horizon, either just before sunrise or just after sunset. The brightness of a comet is, however, directly proportional to the probability that it has already been found by someone else! If you have any ideas about getting a comet named after yourself, a good part of your time (perhaps most of it) should be spent looking in parts of the sky farther away from the Sun. It will be longer before you find a comet that way, but the chances will be better that you are the first to see it.

A comet can appear at any time in any part of the sky. In general, the dimmest comets will be found near the celestial meridian in the middle of the night. The brightness of a comet depends on the size of the nucleus and its proximity to our plan-

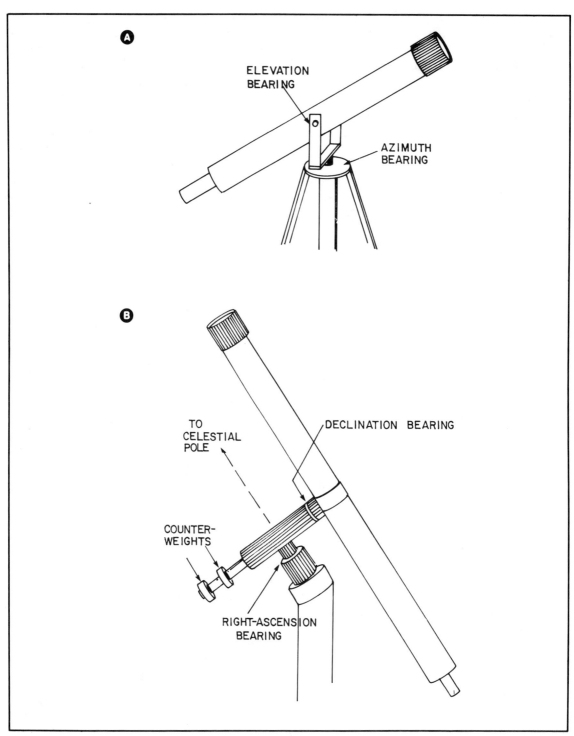

Fig. 4-9. At A, an azimuth-elevation mounting. At B, an equatorial mounting.

et. It is entirely possible to find a reasonably bright comet, for example, near the pole star or in Cassiopeia or in some other part of the sky well away from the Sun.

There is another reason why looking at parts of the sky away from the horizon might yield results. The smaller the angle of elevation of an object in the sky, the more atmosphere its light must pass through in order to reach a telescope at the surface of the Earth. The light from an object at an angle of 30 degrees, for example, must go through twice as much air as the light from an object directly overhead. For an object at an elevation of just 10 degrees, the light must go through more than five times as much air as that from an object at the zenith. This effect is shown in Fig. 4-10.

If you have ever engaged in sunbathing, you know intuitively that you won't get a tan at sunset, but in the middle of a summer day you can get badly burned unless you use a sunscreen. If the air were perfectly clear and perfectly still, this effect would not be very great. In reality, our atmosphere is filled with dust, water vapor, and other light-scattering materials. The constant movement of air causes refraction of the light, blurring images of distant celestial objects. A dim, diffuse comet that stands out clearly at an elevation of 60 degrees might be invisible at an elevation of only 10 degrees.

Comets, unlike the planets and asteroids, do not necessarily orbit near the plane of the ecliptic. Halley's Comet, for example, swings around the Sun at a slant of 18 degrees from the ecliptic. Comets can approach the Sun from north or south of the ecliptic, and could have any possible orienta-

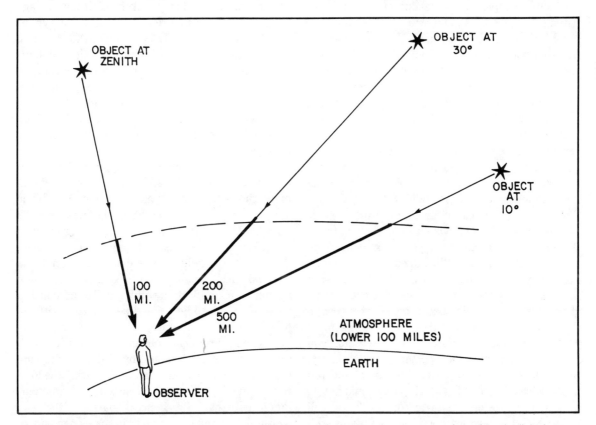

Fig. 4-10. The light from a celestial object at a low altitude must pass through more atmosphere than the light from an object at a high altitude.

tion. There is no reason, therefore, to expect that your chances of finding a comet will have anything to do with whether or not you conduct your search in or near the plane of the ecliptic. (For asteroids, most of which have orbits near the ecliptic plane, the situation is different.)

Comet hunters use different techniques in conducting their searches; some use a zig-zag scheme while others move up and down. Most agree that it is better to concentrate on a relatively small area of the sky, and not to attempt to scan the entire heavens in a single night. Meticulous and repeated scanning is essential. Patience is also needed; it can get monotonous to peer through a telescope or a pair of binoculars for hours and hours.

It has been said that comet hunting requires a high boredom-tolerance level. The boredom can be relieved by listening to a radio; small AM/FM stereo units, complete with headphones for private listening, can be found in most department stores.

Most of the objects you see in the sky are so dim that you sometimes wonder if they are really there. If you come across one of these that looks suspicious, you can avert your eyes slightly and it will appear a little brighter. Experienced comet hunters know about this tip; it has to do with the nature of the retina in your eye. We see color better near the center of the field of vision, near the part of the retina on which light is focused when we look directly at something. Color sensitivity occurs at the expense of the ability of the eye to see dim light. For this reason, far-off or diffuse objects will not be seen if you look straight at them (even if you can see them when you look away from them). The technique of "looking away" or "averting the eyes" is valuable to the astronomer using a small telescope without photographic equipment. It takes a little practice because the eye's resolution is poorer away from the center of vision.

Suppose you do find a celestial object that you believe might actually be a comet. It will probably, although not necessarily, look like a small blurred spot such as shown in Fig. 4-11. You should not go rushing to the telephone or to your desk and call or write all the major astronomical institutions with the exciting news; not right away, at least. First do some things to ascertain, with a reasonable degree of certainty, that you are looking at a previously undiscovered comet.

Detecting Movement

It is entirely possible that you might mistake a nebula or a galaxy for a comet. Both appear as faint, diffuse patches of light as seen through backyard binoculars or telescopes. Nebulae are clouds of gas and dust within our own Milky Way, and they are generally tens, hundreds, or perhaps even thousands of light years distant—well outside the confines of the Solar System. Other galaxies, some of which can be faintly seen with a small telescope, are millions of light years away.

The Great Galaxy in Andromeda, so spectacular when photographed through a large telescope using a long time-exposure, looks like nothing but a dim fuzzy ball through the average small telescope. Figure 4-12A shows a typical nebula that could easily be mistaken for a comet by an uninformed, even if enthusiastic, amateur. Figure 4-12B is a photograph of the nucleus of the Andromeda galaxy. A short time exposure was used to simulate its appearance through a small telescope. In both instances here, the objects look enough like a real comet that confusion is entirely possible.

How do you know whether you are looking at an object inside the Solar System or at something many times as far away as the planet Pluto? The answer can be determined from whether or not the object moves with respect to the background of stars. Planets, asteroids, and comets seem to "wander" across the field of more distant stars; nebulae and galaxies remain fixed. Using a fairly high amount of magnification with a small telescope, the movement of a comet should be apparent within a few hours or possibly less than one hour.

When a promising object is sighted, its position can be precisely noted by making a drawing of the stars in its vicinity, and then plotting the location of the object among those stars. It is important, too, that the general part of the sky (constellation and location within that constellation) be noted because you want to be able to find your

Fig. 4-11. A comet as it might appear through a telescope upon discovery. This is the Comet Mitchell-Jones-Gerber (courtesy of Smithsonian Astrophysical Observatory).

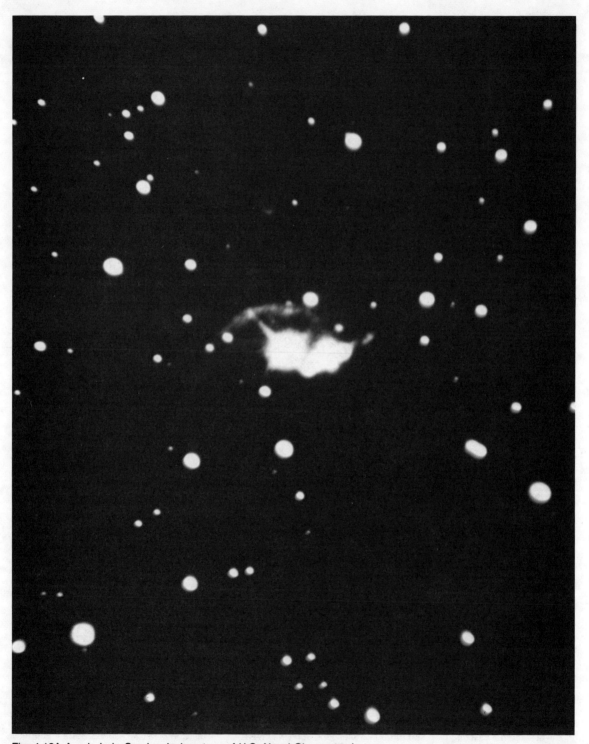

Fig. 4-12A A nebula in Cassiopeia (courtesy of U.S. Naval Observatory).

Fig. 4-12B. The nucleus of the Andromeda galaxy as it might look when seen by the naked eye through a small telescope. Such an object could be mistaken for a comet (courtesy of U.S. Navy Observatory).

prospective comet again! Observations should be made at intervals of about a half hour for several hours. If no movement is detected within that length of time, observations should be repeated the next night.

If the object does move relative to the background of stars, there is a possibility that it is a comet. It could also be an asteroid. This is especially possible if it follows a path close to the ecliptic. The ecliptic, as seen from the Earth, passes roughly through the constellations of the Zodiac: Virgo, Leo, Cancer, Gemini, Taurus, Aries, Pisces, Aquarius, Capricornus, Sagittarius, Scorpius, and Libra.

How do you tell if an object is a comet and not an asteroid? An asteroid always appears as a point of light—even at high magnification. A comet might also look like a point of light, and especially if the comet is far from the Sun. Comets often possess a somewhat fuzzy appearance. This is especially true at high levels of magnification. If the object is well away from the ecliptic, or moves in a direction not parallel to it, then there is an excellent chance that it is indeed a comet.

New or Old Comet?

Suppose that you actually do find a comet after hours of searching the skies. The next question you must ask yourself is, "Have I found a comet that is already known?" There are literally hundreds of these. The serious comet hunter must know about them, and the best way to find out is to contact a professor at a local college or university. Through their astronomy department, you should be able to obtain the latest information. The orbits of the known comets are so widely varied, and there are so many known comets, that such data is completely beyond the scope of this book. Moreover, new comets are found so often that any attempt at presenting such data here would soon be out of date.

After you have done your "homework," if it appears possible that you have detected a comet that no one else has seen before, you should inform the astronomy department at a major college or university. If you are intent (and some are!) on having your name immortalized by being attached to a comet, then it would probably be a good idea to contact a lawyer. While it is doubtful that very many lawsuits will arise over whose name is to be given to a chunk of ice and rock flying through interplanetary space, stranger things have happened.

It is important to realize two things if you happen to come across a comet that you think is "new." First, in any given instance, it probably isn't "new," and you should be emotionally prepared for that near certainty. Second, nobody will ever give you credit for finding a "new" comet unless you let them know about it. You will have to walk a fine line between embarassment on the one hand and a lost opportunity on the other. If you have checked the available astronomical information and have not found a known comet in the part of the sky in which you see your comet, then—and only then—should you get in touch with a major astronomical institution.

Is It Even in Space?

There have been cases in which even serious amateur astronomers have found what they really believed to be comets, only to discover that they were seeing nothing of the sort. If you want to make yourself look and feel foolish, one excellent way is to call some major academic institution on the telephone, tell them about your find, and then discover that you were in fact watching the light reflecting off of a wet spot on a nearby utility pole.

Various earthly phenomena can produce startlingly cometlike "apparitions." The preceding scenario is just one example.

A thin, high cloud such as cirrus, invisible on a dark night, can blur the light from a star or planet in such a way as to create the appearance of a "coma" and "tail." Checking for movement, as described previously, will practically eliminate the chances of your making such a mistake. While the cloud will move within a few minutes, the star or planet will remain fixed for that time relative to the background of stars.

A tiny, high, slow-moving cloud might itself

fool an amateur astronomer into thinking it is a comet. The cloud might gradually change position and reflect the small amount of light it receives from ground-based street lamps. Given two or three hours, almost every cloud will change shape or disappear and give away its true identity.

A large weather balloon, drifting in the upper troposphere or stratosphere and presenting an astonishingly cometlike "coma" and "tail," could fool you. Upon close observation, it will be obvious that the "tail" is not pointing away from the Sun (this would be possible only in daylight). Comet tails always point in a direction generally opposite the Sun.

Satellites, such as the space shuttles, might be mistaken for comets as they reflect sunlight during the hours of twilight. Satellites move rapidly, and much faster than any comet. Within the space of a few minutes, a satellite will move all the way across the sky.

Other natural phenomena, such as Saint Elmo's fire (a glow discharge that sometimes occurs at the pointed ends of lightning rods or radio/television antennas, even in clear weather) can be mistaken for comets. Such things as "swamp gas" could conceivably fool you. And it is not impossible that someday, somehow, a prankster will come up with a way to create a homemade "comet." All comet hunting must be tempered with a little bit of skepticism.

ASTROPHOTOGRAPHY

No matter how small or large the telescope, photography makes it possible to see objects and resolve detail that is invisible with the naked eye. The astronomer at a major observatory, such as Mount Palomar, does not normally sit and peer through the eyepiece of the massive instrument. Instead, the telescope is aimed at a predetermined part of the sky, and a film is exposed at the focal point of the telescope—sometimes for several hours.

Astrophotography is time-consuming, but photographs are far more useful to the professional astronomer than direct visual observation. This is true not only because of the additional sensitivity that can be had with long time-exposures, but because photographs can be studied at length without worrying about changing sky conditions. For these reasons, most of the time at a major observatory is spent exploring very specific, localized areas of the sky, probing mysteries of the universe such as the Crab Nebula, black holes, galaxies, and quasars. Little or no time is given to comet hunting. This is why so many comets are discovered by amateurs. Nevertheless, major observatories frequently find comets by accident on photographs taken for some other purpose.

If you locate a comet and have the equipment to photograph celestial objects, it is worth the time and effort to put the comet on film. This will not only give you a basis for comparison of the nature of the comet from night to night, but it will help you plot its path more precisely than direct visual observation.

Astrophotography is a rather sophisticated art even for the backyard hobbyist with a small telescope. A detailed discussion of astrophotography techniques will not be given here, but the essentials are as follows.

Normally, a telescope of at least 4 inches diameter is used in amateur astrophotography. An aperture of 8 inches is better because the camera shutter time can be cut to one-fourth the value needed with a 4-inch instrument. Other necessities include the following.

A Stable Base. A pedestal type of base is generally better than a tripod.

A Heavy-Duty Equatorial Mounting. The telescope must not shake because of wind. Counterweights must be used to keep the telescope balanced and to minimize torque on the mounting.

A Clock Drive. A clock drive compensates for the rotation of the Earth.

Knowledge of Exact Latitude. For long exposures, the equatorial mount must be adjusted precisely. The latitude should be determined down to the second of arc—and 1 second of arc is a paltry 100 feet north or south—for truly optimum astrophotography.

A Good Camera. A snapshot type of instant

camera won't do. A manual 35-millimeter camera, equipped with a shutter that can be opened indefinitely, is adequate. A shutter cable, that will keep the shutter open without the need for the camera operator to stand by, should be employed.

A Camera Mount. Available for various telescope eyepiece sizes, camera-mount adapters can be found in astronomy hobby shops and some camera stores.

Fast Black-and-White Film. Kodak Plus-X or Tri-X (speeds ISO 125 and ISO 400, respectively) are good choices for black-and-white film.

A Location with Minimum Scattered Light. Good visibility is desirable in any case for the amateur astronomer.

Exposure Time

The exposure meter in the average manual camera will not work in the dim light encountered in photographing stars, galaxies, and comets. Actually, in some cameras, you won't even be able to read the exposure. A certain amount of experimentation (that is, guesswork) will be necessary to find the best exposure time. Furthermore, there is no single "ideal" exposure time when photographing diffuse objects such as nebulae, galaxies, or comets. Different shutter times can produce views that allow observation of different parts of such objects. Shorter exposures reveal internal details. Longer exposures bring out the peripheral details of a diffuse object.

When photographing comets, the exposure time will of course depend on the brightness of the object. In the case of a dim comet that is difficult to see through your telescope, the necessary exposure time will be much longer than for a large comet that looks very bright through the telescope. It might take up to a half hour using Tri-X and more than an hour using Plux-X if the comet is very faint. For a bright comet, the required time might be as short as a minute with Tri-X and two or three minutes with Plus-X. Comets of intermediate brightness would require intermediate exposure times. These estimates might be used as starting points when photographing a comet, but it is important to realize that several exposures will have to be made in order to get the desired results.

The most interesting photographs will naturally result when a telescope or camera is trained on a bright comet. Numerous photos of this kind appear in Chapters 2 and 3. Most of these are photographs of the entire comet, including the tail. The exposure times are sufficiently long so that the tail is clearly visible. There is plenty of room for experimentation with exposure time on the part of the amateur.

Suppose you focus in on the head of a bright comet, and vary the exposure time over a range of say three or four to one. Imagine that you have already found that two minutes produces a good picture. You might then try exposure times as short as one minute or as long as four minutes, "bracketing" your exposure times in geometric proportion. The fastest shutter speeds (shortest exposures) would give a better view of the nucleus, while the slower shutter speeds would allow you to see the outer part of the coma. Using less magnification, the same scheme would let you observe various facets of the tail. The fastest shutter speeds would reveal only the inner part of the tail near the head, while the slowest speeds would bring out the faintest wisps of the tail at greater distances from the head.

Perhaps you are wondering exactly what is meant by "geometric proportion." This means simply that the difference in shutter speed as measured in absolute terms (seconds or minutes) becomes less and less significant as the exposure time increases. It is the percentage difference, not the absolute change, that is important. For example, a 10-percent increase in the shutter speed could be the difference between 60 seconds and 66 seconds (that is, 6 seconds), the difference between 10 and 11 minutes (that's a full minute), or the difference between 60 and 66 minutes (that's 6 minutes). The film does not care about the absolute change in shutter speed, but about the percentage difference. Changing the exposure time by one second would make a lot of difference if the original exposure was two seconds; it would make no difference if the original exposure was two hours.

A geometric progression over a four-to-one range, where t represents the shortest exposure time, is shown in A of Fig. 4-13. A six-to-one range is shown in B of Fig. 4-13, and a ten-to-one range is shown in C of Fig. 4-13. You should choose points that are equally spaced along the horizontal axis (examples are given in the graphs: five points at A, five points at B, and nine points at C). The "center" or "ideal" speed is 2t at A, 2.5t at B, and 3.2t at C, representing the geometric mean of the minimum time t and the maximum time 4t, 6t, or 10t.

While you do not have to use the geometric-progression technique when varying the exposure, it could save you some time and film that might otherwise be needlessly spent at the slower shutter speeds.

Making the Prints

For truly high-quality astrophotography, you should get acquainted with the people at the best photo lab you can find in your area. This should not be too hard if you submit a lot of photos for 8-x-10-inch prints; they'll learn to love you in a hurry after you spend a hundred dollars on them. The so-called "custom" prints are pretty expensive. They can cost five or six dollars apiece as compared with about two dollars for "machine" prints. The difference in quality is worth every penny of the difference in cost.

When you submit a roll of film for developing, you should first have the negatives printed on a proof sheet. The proof sheet will tell you which exposures are suitable for making into prints. The proof sheet and developing for a 20-exposure roll of film will cost you about five dollars.

When specifying a print, be sure that the lab makes no mistake about it. The negatives are numbered sequentially: 1, 1A, 2, 2A, 3, and so on. Make certain they know which exposures you want made into prints. Generally, you will want to get 8-x-10 custom prints. The people at the photo lab will advise you about the different border styles and finishes from which you can choose.

The Problem of Odd Movement

With long-exposure photographing of comets, a peculiar problem arises for the amateur. In a time period as short as a few minutes, a comet will move significantly with respect to the background of stars. Clock-drive units for amateur use will follow the stars perfectly, turning at the rate of 15 degrees per hour from east to west along the ecliptic. The stars will appear as bright points of light even if the exposure time is very long. This is assuming the clock drive is centered precisely on the North Celestial Pole (or the South Celestial Pole in the Southern Hemisphere). The comet will appear blurred because of its motion. Depending on the movement of the comet, the blurring might occur at any angle. To prevent this, it would be necessary to make the telescope follow the comet instead of the stars, and ordinary clock drives cannot generally be made to do this.

When the camera drive is made to follow the comet, the result is that the stars trace out short arcs instead of appearing as points of light. If the clock drive is set to follow the stars, the comet would appear "smeared" sideways on the photographic plate, and all of the detail would be lost.

Unfortunately, there is no simple solution to this problem for the amateur who intends to photograph dim comets. If the progress of the comet across the sky is exactly 15 degrees per hour—an extremely unlikely coincidence at any given time—then the equatorial mount can be turned so that the clock drive will follow the comet instead of the stars.

ASTROPHOTOGRAPHIC COMET SEARCHING

The odd-motion problem can be used to advantage in searching for comets. This is especially true for distant or small comets that might be overlooked during direct visual searches because of their resemblance to stars.

Suppose that you have a telescope with a short focal length at relatively low magnification, a clock drive, and the other astrophotographic equipment described previously. Further suppose that you simply point your telescope at some point in the sky,

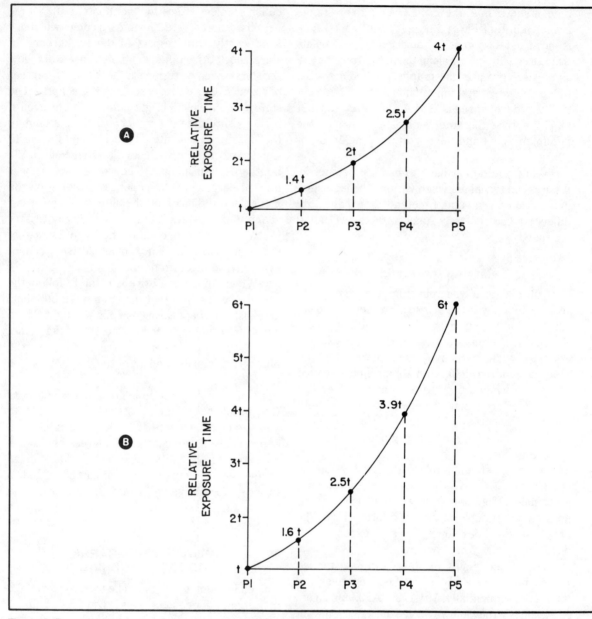

Fig. 4-13. Examples of geometric relationships for camera exposures. At A, four-to-one; at B, six-to-one; at C, ten-to-one. Exposures are denoted by P1 through P5 (A and B) and P1 through P9 (C).

open the camera shutter and—using slow black-and-white film such as Kodak Panatomic-X—expose the film for a very long time (perhaps two hours).

Assuming conditions are ideal so that background (scattered) light is minimal, you can take many beautiful photographs of the Milky Way, nebulae, and other galaxies in this way. You might also find a comet. Because of its motion with respect to the background of stars, the comet will trace out a short line on the film (Fig. 4-14). If the

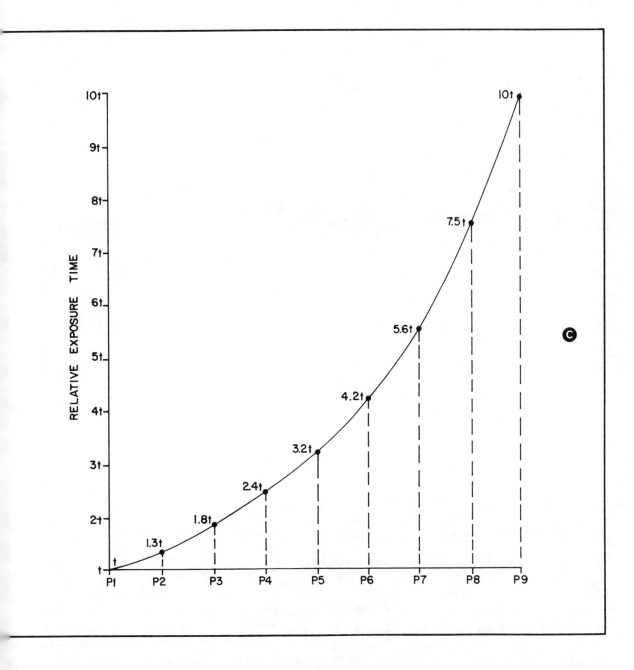

line is near and parallel with the ecliptic, the object is probably an asteroid. If the line is not near the ecliptic or is not parallel with it, the chances are excellent that the object is a distant comet.

A special advantage of slower film and long exposure time is that, usually, the detail is better than with faster film and shorter exposure times. If the equatorial mount is aligned precisely, a Panatomic-X negative can have enough detail to justify looking at it with a low-power microscope using 30 to

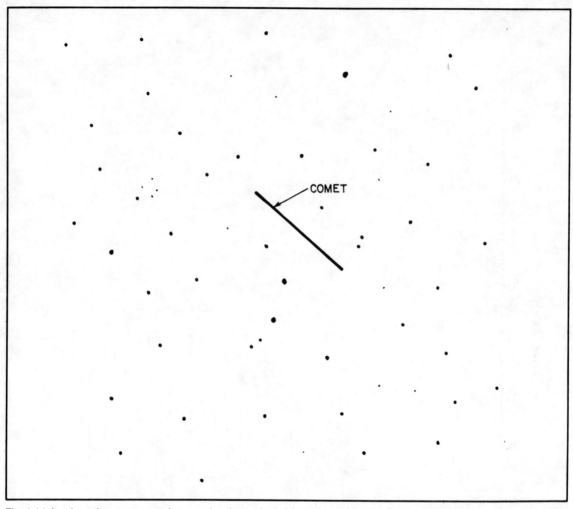

Fig. 4-14. In a long time-exposure photograph using a clock drive to keep the stars fixed, a comet would show up as a line.

40 magnifications. A black-and-white negative can also be cut and mounted in a 35-millimeter slide holder for viewing with a slide projector.

On the negative, the stars and other celestial objects will appear black on a white background. Astronomers often use negatives, rather than prints, to look for detail in photographs of the heavens because the negative is a "first-generation" exposure and the detail is not limited by the quality of photographic paper used in making a print.

The chief advantage of the photographic-search method over visual comet hunting is the lesser likelihood of missing tiny, dim comet. The photographic method is, however, much more time consuming because smaller regions of the sky are covered per unit time. Nevertheless, a systematic, night-after-night photographic search might prove to turn up more truly new comets than a visual hunt where so many old comets are rediscovered. For those amateurs willing to go to the trouble (and expense) of a photographic search, the investment could provide fruitful. Keep in mind that comet hunting is a sport in which there is no guarantee of success. This discussion should not be taken as a recommendation that a photographic search should be conducted or that it is better than a visual hunt, but only as an opinion that a photographic search might be worthwhile.

TRACKING

Once a new comet has been discovered, professional astronomers give it a name and make observations for the purpose of calculating its orbit. Precise determination of the orbit is no easy task; the margin for error in observation is almost vanishingly small. A tiny discrepancy in the observations can make a large difference in the predicted perihelion distance and time of passage as well as the length of the period. Actually a nonperiodic comet might be thought at first to be periodic, or a periodic comet might appear to be nonperiodic. As observations continue, the precision improves until the perihelion distance, time of perihelion, and length of period are known almost exactly.

The problem of orbit calculation is compounded by the fact that comet orbits do not generally coincide with the ecliptic plane. Astronomers are thus faced with a three-dimensional problem. The calculations would take years with a pencil and paper or days with an ordinary calculator. Computers are used to carry out the arithmetic and to provide graphics displays of the projected path of the comet.

Basically, astronomers try to answer these questions:

☐ What will the perihelion distance be?
☐ When will perihelion occur?
☐ Is the orbit periodic, and if so, how long is the period?
☐ What will be the comet's day-to-day path through the heavens as seen from Earth?
☐ How close will the comet come to the Earth?
☐ When will viewing conditions be best in various parts of the world?

As technology improves, astronomers will try to unravel further mysteries about comets:

☐ Do they exhibit spin, and if so, is any spin orientation or rate especially common?
☐ Do comets have magnetic fields, and can this affect their orbits?

Usually, we can get a pretty good idea of what kind of comet we have found well before it reaches and passes perihelion. As we have seen in the cases of Comets Kohoutek and Comet West, sometimes there are surprises that cannot be predicted.

WHY SEARCH?

After reading this chapter, some people will certainly be wondering why anyone would want to spend hours and hours looking for little pieces of ice and rock that tumble through interplanetary space. They will say that if a comet appears, we'll notice by and by; leave the discoveries up to the professionals.

As we have seen, however, the big observatories just don't have the time to conduct comet hunts. It is a group of devoted amateurs (or professionals in their spare time) who engage in this activity, and who make a sizable proportion of the discoveries. In terms of personal satisfaction, the reward of finding a new comet is akin to the elation that any scientist experiences when he or she makes a breakthrough.

There is one more compelling reason to look for comets that amounts to a virtual mandate. It is possible for a comet, especially a small or dim one, to pass unnoticed right up until it smashes into our planet! This could happen if a comet came around the Sun in such a way that it was always hidden in daylight. We would never see the comet unless it was unusually bright.

A hypothetical situation of this kind is shown in Fig. 4-15. The comet would be visible through telescopes well before perihelion, but the major observatories—investigating small regions of the sky—might miss it. After perihelion, the comet might again be visible near the horizon immediately before sunrise or after sunset—if a telescope were used for the purpose of comet hunting. The comet could be very hard to find even with meticulous scouring of the sky.

The Tunguska Event of 1908, marked by an

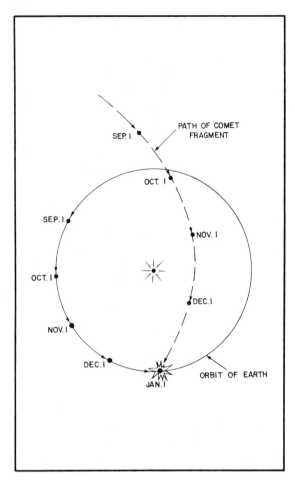

Fig. 4-15. A small, dim comet might remain invisible right up until it hit the Earth. This is especially possible if the comet maintained a position in the sky close to the Sun near perihelion.

explosion comparable to a nuclear detonation, is thought to have resulted from the impact of a comet fragment in Siberia. If this was a comet or comet fragment, why didn't people notice it in the sky before it hit? Researchers have found that the object would have appeared very close to the Sun in the heavens, and thus it would have been invisible to anyone except a comet sleuth who might be looking for a tiny, dim comet near the horizon just before sunrise or just after sunset.

The Tunguska comet fragment (assuming that is in fact what it was) was probably a telescopic object even as it bore down on our planet. No one would have been able to see it with naked eyes. Yet the consequences of the impact were devastating. Trees were leveled for miles around "ground zero." The shock wave caused serious damage up to a hundred miles away. The atmospheric shock wave circled the globe before dying out completely.

The Tunguska Event took place in a sparsely populated part of the world. But that was just a coincidence. If the object had landed just a few hours later, it could have struck Leningrad, Stockholm, Oslo, or Anchorage.

How often do comet fragments actually strike the Earth? We do not really have much of an idea. In 1978, there were over 4 billion people in the world. In 1908, the world population was a little more than 1 1/2 billion. In 1800, there were less than 1 billion people on our planet. At the height of the Roman Empire, there were just 300 million. Comet and meteorite impacts, unless cataclysmic, might have passed unnoticed more often than not in past centuries when scientists did not have sophisticated seismic and barographic equipment and when communication and transportation were slower and less reliable.

We do know that a major bolide strike has not occurred since the 1908 event. That tells us little. The next impact might happen a thousand years from now, a century from now—or a week from now.

Our technology today makes it conceivable that, if a comet fragment were detected on a collision course with our planet, we could blow it up or divert it before the catastrophe. But not unless we knew it was there!

This might sound farfetched, but maybe someday an amateur comet hunter, peering through a telescope at some part of the sky from a hilltop, will find a comet that has not been previously catalogued, will notify the observatories, and ultimately be recognized as having saved thousands or millions of people from death at the hands of a chunk of ice and rock from space.

Chapter 5

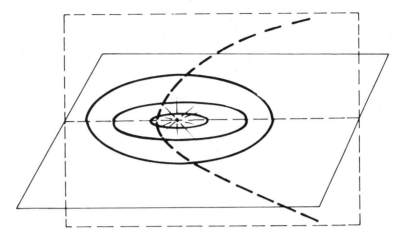

Stones from Space

AN INTENSE METEOR SHOWER IS AN UNFORgettable spectacle. So is a single large meteor or meteorite as it lights up the sky for a brief moment and leaves a glowing trail. The Indians called them shooting stars. Other ancients also seem to have had the idea that meteors were stars that fell from their places in the sky, something like angels cast out of heaven. Perhaps stars "fell" when they had committed some sort of sin or when they got old and died. Careful observation of the sky would have revealed that there were no stars missing after the appearance of a meteor or meteor shower.

Today, we know that meteors are small pieces of interplanetary debris that get swept up by the Earth's gravitation and pulled into the atmosphere. The heat of friction with the atmosphere causes the meteors to glow. Usually they burn up completely before they hit the ground, but occasionally a large meteor does reach the surface. Then we call it a meteorite. Some people have actually seen meteorites fall and discovered them lying on the ground, still hot. Meteorite impacts are, fortunately, rare enough so that people are not individually threatened by them. You do not have to live in constant fear that a basketball-sized piece of space rock will come down upon your head.

Massive meteorites, perhaps more appropriately called asteroids, have struck the Earth many times since our planet was born 4 1/2 billion years ago. Some meteorites were no larger than ordinary stones when they reached the ground, some were veritable boulders, and a few were thousands of feet or even several miles in diameter. Such large meteorites, while rare nowadays, can and do still fall from time to time. Today we have the technology to detect massive meteoroids before they fall. We are on the verge of acquiring the ability to divert them or blow them up before they can deal a blow to us.

SPACE DEBRIS

When the Solar System was much younger, and the planets had formed but had not yet reached their present stable states, there was probably much more debris in interplanetary space than is the case today. The most common theory of the origin of the Solar System holds that the planets condensed out of rings of gas and dust revolving around the newborn Sun. This theory and others are discussed in Chapter 1.

Gradually, the material in each ring (with the exception of the ring between Mars and Jupiter) pulled itself together in a concentrated sphere because of mutual gravitation among the particles. The result was the formation of planets at various distances from the parent star. Some of the planets, especially the larger ones, acquired smaller daughter orbs. Mercury and Venus did not develop moons, but Earth and all of the rest of the planets did.

Although most of the material in the planetary rings found its way into the planets themselves, some material remained free. Some of the original debris that made up the protoplanets is still orbiting the Sun in rings. In a sense, the process of planet formation is still going on (although we can say that it is complete for all practical purposes). Still, fine particles of space dust still rain down on all of the planets and their moons. Among these fine particles are larger fragments. Some objects are as big as pebbles, a few are the size of baseballs or footballs, and a very small minority are as large as cars, houses, or whole cities.

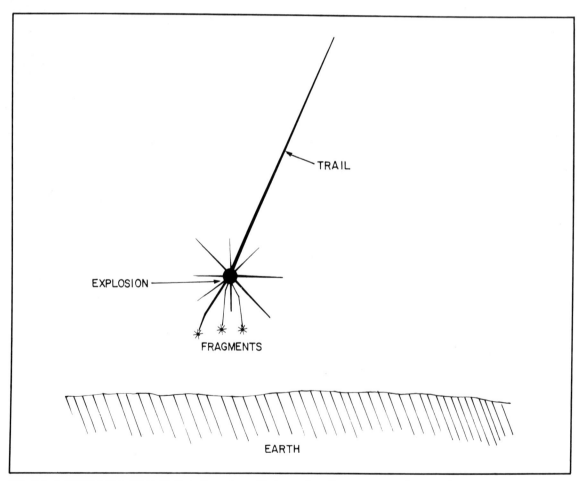

Fig. 5-1. A drawing of a meteorite fireball, as it would appear in a time-exposure photograph.

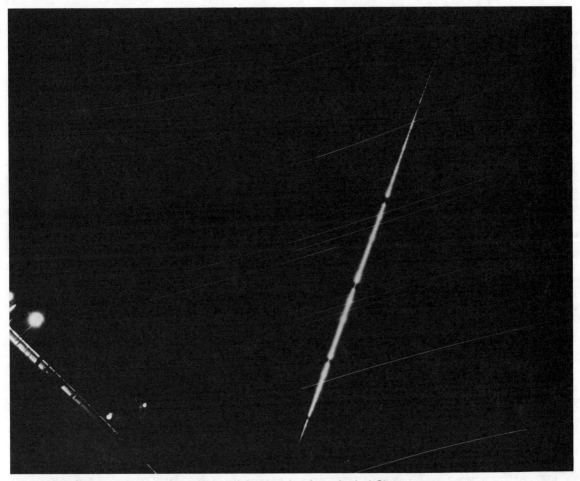

Fig. 5-2. Photograph of a meteorite (courtesy of Smithsonian Astrophysical Observatory).

The planetary rings are not the only source of space debris. Comets also contribute to meteor activity. All comets eventually disintegrate as their icy material evaporates from the radiation of the Sun on repeated perihelion passages. When a comet loses all of its ices, only a swarm of small meteoroids is left. This swarm follows essentially the same elongated orbit of the original comet, but becomes more and more diffuse and spread out along the orbit.

When the Earth passes through or near this swarm, we see a flurry of meteors that astronomers call a meteor shower. Most of these meteors appear as brief flickers of light that rapidly move across a small span of the sky. Some meteors burn up and occasionally explode into brilliant fireballs (Fig. 5-1). If the fireball reaches the Earth (Fig. 5-2), one or more meteorites will be found at the site of impact. A large meteorite will cause an explosion, leaving one or more craters, when it hits. Some large meteorites explode in midair because of the heat produced as they encounter the atmosphere. When this happens, dozens or even hundreds of fragments will be scattered over several square miles of the Earth's surface (Fig. 5-3).

Not all meteorites are witnessed by people. Even today, with billions of people on our planet, the vast majority of meteorites fall in sparsely populated regions. Moreover, three-quarters of them will never be found on land. The Earth is

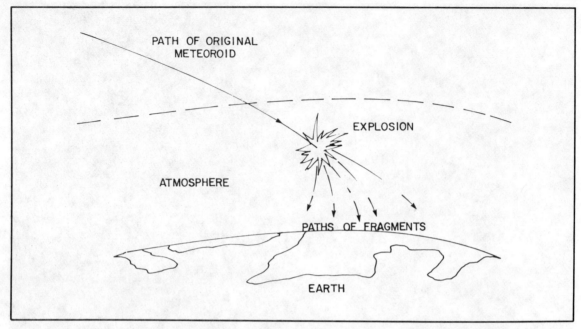

Fig. 5-3. A large meteoroid might explode as it heats up in the atmosphere, producing several small meteorites that land over a large area.

largely covered by oceans and lakes, and most meteorites splash down in these bodies of water and rapidly sink to the bottom.

Meteor showers and fireballs have occasionally been classified as unidentified flying objects. An uniformed person, driving a car along a highway at night, might witness the breakup of a meteorite and get the impression of a swarm of fast-flying craft. Often, the fireballs change color or appear in different colors as they heat up and cool off. In rare instances, dozens or even hundreds of people witness the event, and it makes the news. In August, 1984, people in the Pacific Northwest, the Northeast, and the South reported strange lights seeming to move from north to south through the atmosphere. I saw a brilliant fireball in the early morning; it left a trail and was obviously a meteor. Astronomers attributed the sightings to the Perseids, which is one of the most spectacular annual meteor showers.

CRATERS

Craters provide vivid evidence of past meteorite activity on planets having little or no atmosphere. The Moon is a good example where many craters can be seen even through a small pair of binoculars. A good amateur telescope reveals thousands of craters on the Moon.

So many craters can be seen on Mercury that the surface of that planet resembles the Moon (Figs. 5-4A and 5-4B). Mars has many craters that provide evidence of past bombardment by meteorites. The moons of the outer planets also are peppered with craters. Phobos, one of the moons of Mars, has craters so large in comparison to its size that one wonders why the moon didn't shatter from the impacts (Fig. 5-5). Mimas, one of Saturn's moons, has a formidable scar that dominates its surface. In addition, there are numerous smaller craters (Fig. 5-6) on Mimas. Enceiadus, another of Saturn's moons, is also covered with craters (Fig. 5-7).

We do not see very many craters on the surface of the Earth because the atmosphere has eroded most of them beyond recognition over thousands or millions of years. No craters have been seen, even by the space probes, on Venus, Jupiter, or

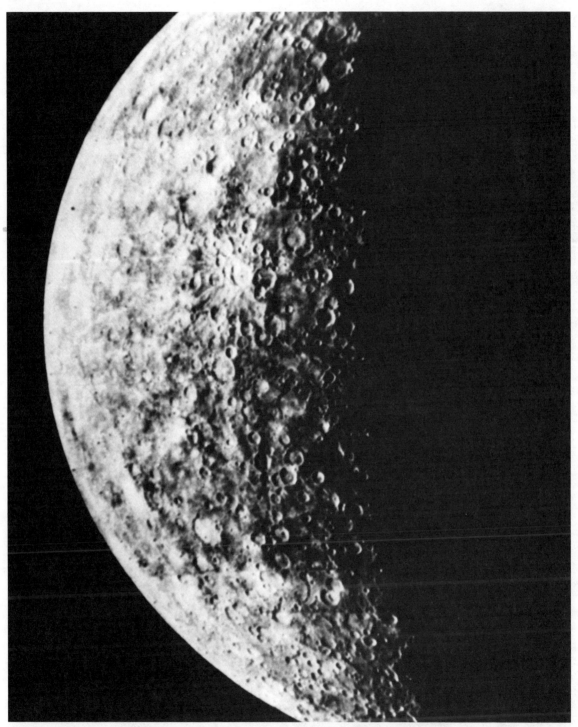

Fig. 5-4A. The cratered surface of Mercury. The crescent phase as seen from several thousand miles out in space (courtesy of NASA).

Fig. 5-4B. A closeup view, showing crater details (courtesy of NASA).

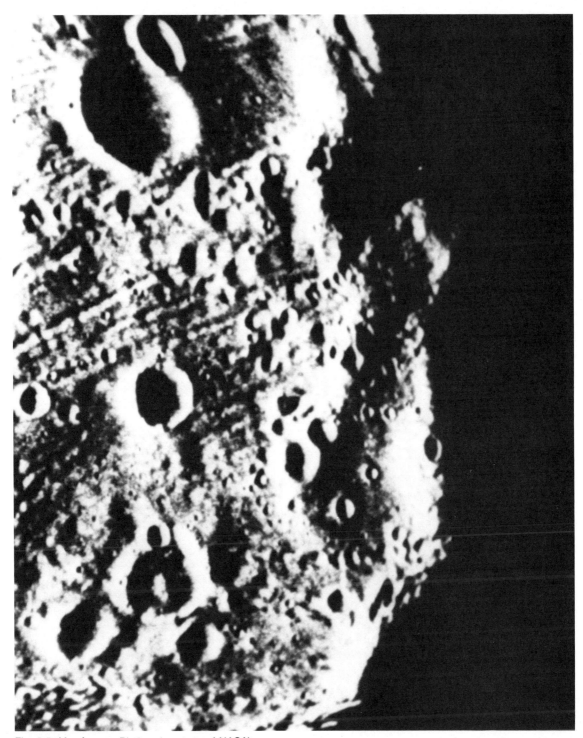
Fig. 5-5. Mars' moon Phobos (courtesy of NASA).

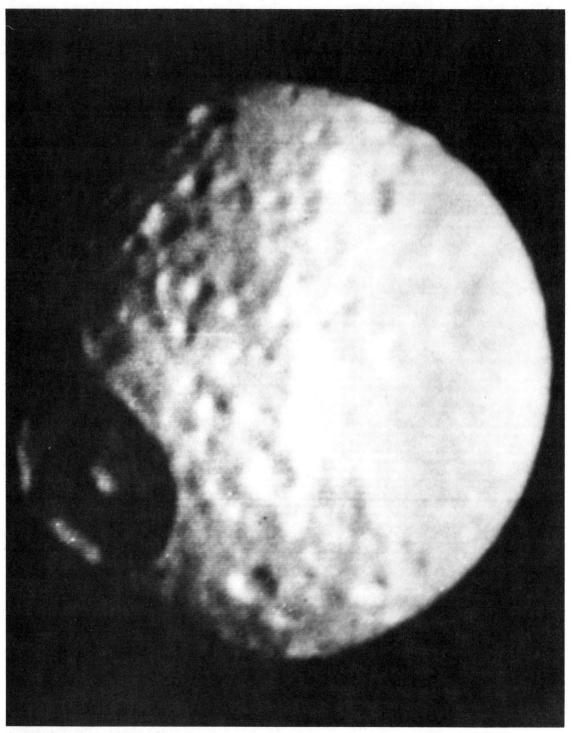

Fig. 5-6. Saturn's moon Mimas shows how large a crater can be with respect to a world (courtesy of NASA).

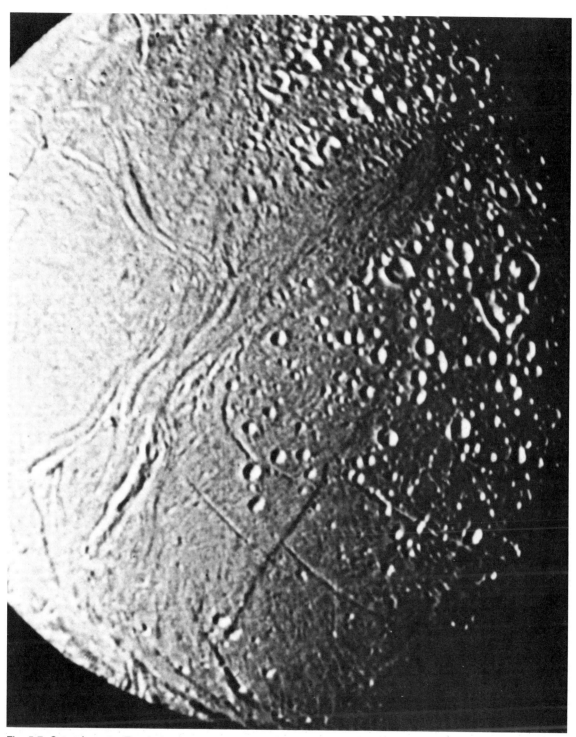

Fig. 5-7. Saturn's moon Enceladus (courtesy of NASA).

Saturn. This can be attributed to the thick clouds that cover these planets. Venus, having a solid surface, probably does have some craters, but erosion has probably all but obliterated most of them. Jupiter and Saturn might not have definable surfaces or they might be covered by strange liquid-hydrogen oceans. Therefore it is possible that there are no craters on those planets. The same could well be true for Uranus and Neptune. No closeup pictures have yet been obtained for Pluto and its moon Charon, but they are probably covered with craters (having an appearance ironically similar to the broiling hot Mercury).

Meteorite craters tend to be much larger than the objects that make them. This is because of the tremendous force that accompanies the crash landing of a large, solid object at a speed of tens of miles per second! You can create your own craters by throwing small stones into sand and see for yourself. The size and configuration of a crater depends on the speed of impact, the size of the meteorite, the angle of impact, and the composition of the surface material.

Large meteorites form craters with special features. One of these is the formation of a central "minimountain" at the base of the crater near the point of impact. Many of the craters on Mercury (Fig. 5-4B) have this feature. The large crater on Mimas (Fig. 5-6) has a prominent central peak. Many of the larger Lunar craters have hills or mountains at their centers.

Another feature of large craters, especially visible on the surface of the Moon, are radial markings—known as rays—that extend outward for hundreds or thousands of miles from the point of impact. Rays are believed to be the result of debris thrown outward by the force of the meteorite impact. Some meteorites, called tektites, have been found on the Earth and are thought to have come from the Moon after a huge meteorite landed there.

Some meteorites cause fractures in the surface of the Moon or planet on which they land. Sometimes such fractures take the shape of concentric rings around the point of impact (Fig. 5-8). In other cases, there is no particular pattern as the surface cracks along weak points near the point of impact.

Especially massive meteorites can result in volcanic activity that greatly alters the surface of a moon or planet. The maria ("seas") of the Moon were probably formed by volcanic lava erupting as a result of severe meteorite impacts (although it is possible that some Lunar volcanoes were not started by meteorites). A large meteorite landing could increase volcanic activity not only near the point of impact, but all over the whole moon or planet.

Small craters cause little disturbance on a planet, and appear like concave indentations of various sizes. On moons or planets having little or no atmosphere—such as our own Moon—tiny meteoroids reach the surface without breaking up, and they can form craters just a few feet or inches across. The Apollo astronauts encountered these; in Fig. 5-9, a tiny crater can be seen behind Edwin Aldrin of the Apollo 11 mission. The crater is at the left and measures about 3 feet across. In the background another similar-sized crater is visible.

HOMEMADE CRATERS

You can make your own small craters on an experimental basis using some flour or corn starch, small dried beans or pellets, and a drinking straw. Other methods include throwing golf balls or baseballs or rocks at sand or snow. Homemade craters can be made to greatly resemble the small and medium-sized craters that pepper the surface of the Moon.

Put several cups of fine flour or corn starch into a deep pan such as a flat cake pan. The flour should be at least an inch deep. Then put the pan on the floor and put a bean in the straw, stand 1 or 2 feet away, and blow the bean at the center of the pan.

Don't choke on the bean, and be sure no one is standing in a place where they might get hit. If you are worried about bugs getting at the flour or corn starch—some of it will fly around upon impact—conduct the experiment outdoors.

If you happen to live in a place where dry, pow-

Fig. 5-8. An impact basin on Callisto, one of the larger moons of Jupiter (courtesy of NASA).

Fig. 5-9. Some craters are just a few feet across, such as the one behind Edwin Aldrin in this Apollo photograph (courtesy of NASA).

dery snow is on the ground, you can conduct this same experiment using a baseball or softball. Just throw the ball at the snow. Wet snow won't work well—the ball will simply cut a hole in it—but very dry snow can produce some interesting craters. (If you lose the ball in the snow, don't worry; you'll probably find it in the spring when the snow melts.)

If you live near an ocean beach that has fine sand, you can throw rocks or baseballs or golf balls at the sand and watch the resulting craters form. Again, be sure you don't hit anyone with whatever you throw. The "meteorite" might bounce off the sand fly back up into the air. This is especially likely if the angle of impact is small.

Perhaps you will want to photograph your craters. For best results, a point source of light—such as an unfrosted incandescent bulb or sunlight on a cloudless day—should be used. The lighting angle should be low so that the illumination falls on the "surface" at a sharp slant, casting long shadows on any irregularities. The light source should come from the right or left of the camera (not from behind it or in front of it).

The craters and mountains on the Moon or the planets stand out most vividly near the terminator, where the Sun is low in the sky. Craters and mountains in a pan of flour, corn starch, or in the snow are no different. It is important that the light falls on the surface at an angle of 20 degrees or less (Fig. 5-10). The camera should be set for a slightly faster shutter speed (shorter exposure time) than the light meter would indicate so that the contrast on the bright white surface will be emphasized.

It is interesting to experiment with various angles of impact, different "meteorite" speeds, different "meteorite" densities, and different surfaces. Figures 5-11A through 5-11D show several examples of artificial craters. In each case, note how debris is scattered around the rims of the craters. If the angle of impact is small, debris is thrown farthest in the direction opposite the arrival of the "meteorite," and the crater itself is elongated. If the "meteorite" comes in practically straight down, the debris will be scattered uniformly around the edges of the crater, and the crater will be nearly circular in shape.

CRATERS ON THE EARTH

When you look at the Moon through a telescope and see thousands of craters, you must wonder why our own planet, so near to the Moon in space, is not also covered with craters. The reason is erosion; the constant movement of air and water wears the evidence away. A million-year-old crater on the Moon looks like a crater formed on the Earth just a generation ago because there is no air on the Moon to erase craters.

The lack of erosion on the Moon results in some haunting effects. The footprints of the men who first walked on the Moon will still be there, prac-

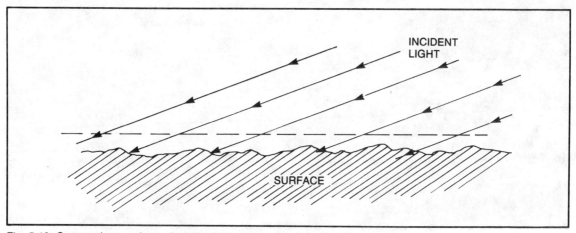

Fig. 5-10. Craters show up best when the light strikes a surface at a sharp angle.

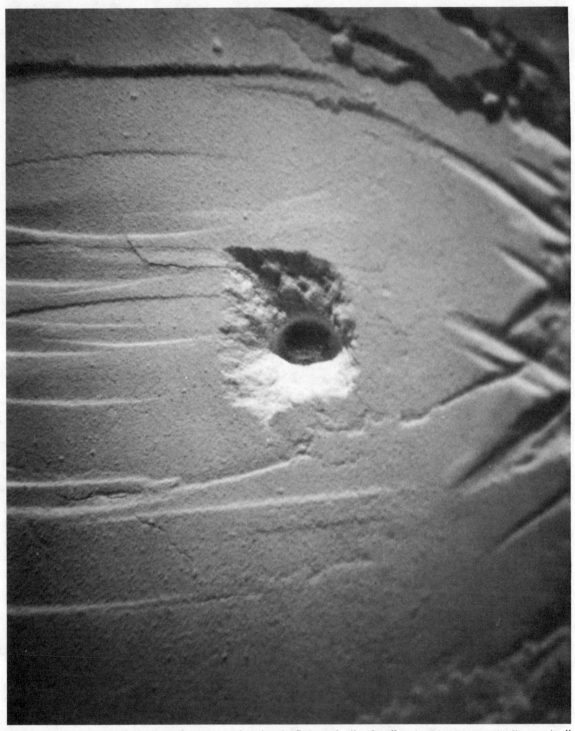
Fig. 5-11A. An example of a homemade crater made using dry flour as the "surface" and a dried bean as the "meteorite."

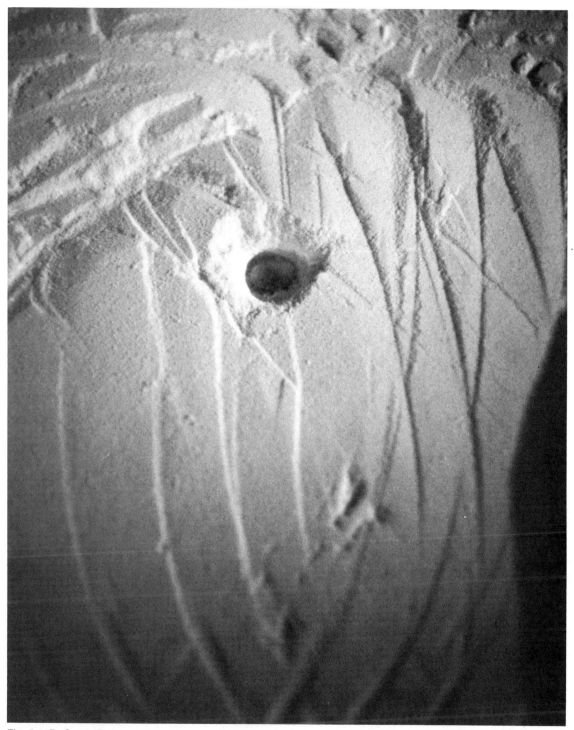
Fig. 5-11B. Crater B.

Fig. 5-11C. Crater C.

Dione, a moon of Saturn, is typical of planets and satellite with thin or nonexistent atmosphere. Dione is covered with craters, the scars from cataclysms of long ago. The interplanetary realm is not a complete void; it is filled with fragments ranging in size from mere grains of dust to massive asteroids hundreds of miles across. Dione, along with other planetary moons, bears testimony to this. (NASA photograph.)

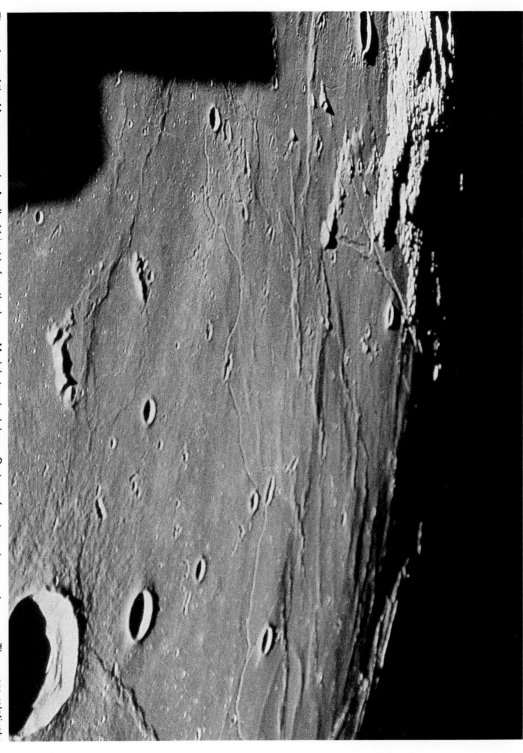

The surface of the Moon as seen by Apollo 11 just before the Lunar Module touched down. Craters of various sizes can be seen. These are relatively new craters, formed in a region of the Moon recently modified by volcanic activity. Perhaps a huge meteoroid impacted in the area just a few million years ago, precipitating a volcanic eruption, and erasing all evidence of previous catastrophes. (NASA photograph.)

Fig. 5-11D. Crater D.

tically unchanged, in a thousand years. Some evidence of them will probably still be left in a million years, when our species (if we still exist) is radically different than it is today. Small craters remain on the Moon for millions of years; large ones last for eons. When you look at the Moon through a pair of binoculars or a small telescope, you are looking at the work of meteorites. The imprints are a historical record extending back in time all the way to the origin of our Solar System.

The Earth has been struck more often by large meteorites than the Moon. This is because the Earth has about seven times the surface area of its sister planet. Also, the gravitation of the Earth, being stronger than that of the Moon, pulls harder at stray debris in space and increases the chances that meteoroids will come down to the surface. But a brief glance at any satellite picture of the Earth fails to reveal an abundance of craters.

If most of the meteorite landings took place when the Earth was young—less than a billion (1,000,000,000) years old—then we should not be surprised that all the evidence has been lost. Ocean waves crashed incessantly onto the rims of those craters, for days, years, millennia, and eons. The frequency of great meteorite catastrophes steadily diminished until, nowadays, we think of a cosmic disaster as impossible. That is almost true, but not quite.

There are craters on the Earth. Probably the most famous is the Barringer crater in Arizona. It looks as new as any crater on the Moon. This crater, which is about a mile in diameter, was formed a few thousand years ago. The impact must have been utterly catastrophic, causing an explosion more powerful than the detonation of any hydrogen bomb man has yet devised. The Earth must have shook for hundreds of miles around the impact site; perhaps earthquakes occurred all over the world. It is likely that huge tsunamis circled the globe, inundating the coastal regions of such places as Hawaii, Japan, and eastern Australia.

While a crater one mile across seems large to us, the Barringer crater is by no means the largest one on our planet. It is believed that the Hudson Bay basin could be a huge meteorite crater (Fig. 5-12). It must have formed many millions of years ago. The violence of the impact, caused by an asteroid perhaps a half mile in diameter, would probably exterminate humanity if it occurred today.

Many lakes appear to be meteorite craters. Crater lakes, as such lakes are called, are characterized by an unusually symmetrically round or oval shape. They are generally deeper than ordinary lakes in proportion to their size. The water in a crater lake is often colder than the water in ordinary lakes nearby. Crater lakes form when a meteorite lands in an area with abundant ground water so that the crater bottom is lower than the water table.

Some Earth craters can be seen from space. Irregularities in the terrain can often be ascribed to past meteorite impacts (as shown in Fig. 5-13). Any circular or elliptical feature might be suspected as a meteorite crater.

Meteorite craters should not be confused with craters of volcanic origin. Volcanic craters are always found at the tops of mountains; a good example is the huge crater at the top of Mauna Loa on the island of Hawaii. A meteorite crater might be found anywhere.

Craters tend to last long in locations characterized by little rainfall and sparse vegetation. This is because rain accelerates the erosion process, and vegetation modifies the landscape. In jungles, forests, or windswept areas, craters are covered up quickly by vegetation or dust, and erosion takes a much shorter time. Craters will be more visible in some parts of the world than in others even though meteorites have fallen all over our planet.

Crater erosion takes place on other planets (not just on Earth). Any planet or moon with an atmosphere will have shorter-lived craters than a planet or moon without an atmosphere. In the Solar System, Venus and Mars, as well as Earth, have solid, identifiable surfaces, on which meteorites can fall, and substantial atmospheres that cause general erosion of the surface.

We have not yet obtained general, detailed views of the surface of Venus. When we do, it is likely we will see evidence of craters. On Venus, craters would probably erode very fast because the

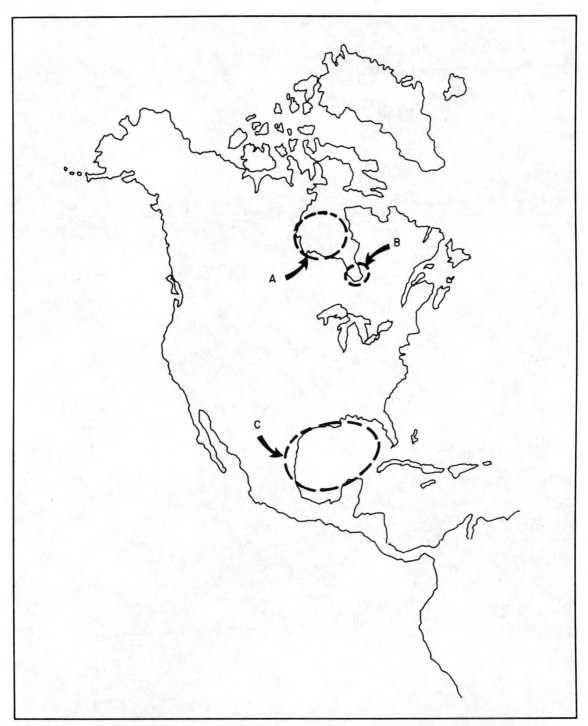

Fig. 5-12. Some geographical features of our planet might be the remains of old crater basins formed by massive meteorites. Notable examples in the Western Hemisphere are Hudson Bay (A), James Bay (B), and the Gulf of Mexico (C).

Fig. 5-13. A photograph of the Grand Canyon and surrounding terrain, taken from several hundred miles up. Irregularities at center and top show evidence of past meteorite impacts (courtesy of NASA).

atmosphere is extremely dense. Mars is another planet that is shrouded by an atmosphere—albeit a thin one—and craters on that planet are plainly visible when dust storms are not raging.

Violent winds, sometimes at speeds of 300 miles per hour, are known to blow on Mars, and the reddish sand is whipped high above the surface and carried along for hundreds or thousands of miles. Erosion is evident around some of the craters on the red planet (Fig. 5-14).

The thicker the atmosphere of a planet or moon the larger a meteor must be in order to hit the ground and form a crater. If a small fragment does reach the ground on such a planet or moon, the tiny crater will be swallowed up rapidly by the environment, and all traces will disappear within a few years or decades. This is why we do not find tiny craters on our planet (as the Apollo astronauts found when they landed on the Moon). The minimum size of craters on a planet or Moon is directly proportional to the thickness of the atmosphere. We should expect to find only huge craters on Venus, moderate-sized to large ones on the Earth, fairly small to large ones on Mars, quite small to large ones on Mercury, and craters of all sizes on our Moon and the smaller moons of other planets.

The larger moons of Jupiter and Saturn have atmospheres, and it is reasonable to suppose that craters on those satellites have eroded in a manner similar to those on Venus, Earth, and Mars. Titan is the most interesting of the planetary moons. In orbit around Saturn, Titan attracts the attention of astronomers because of its abundance of chemicals that are believed to have given rise to life on our own planet. The atmosphere of Titan is an opaque orange-yellow. Clouds make it impossible to resolve details of the surface, even as seen from nearby space probes. In this respect Titan is a little like Venus. When we finally send a landing probe to Titan, we will probably find eroded craters similar to the ones on Earth or Mars.

METEOR SHOWERS

If you look at the sky for a long enough time on any given night, sooner or later you will see a meteor. Meteors usually look like silver- or gold-colored streaks of light lasting from a fraction of a second to perhaps two or three seconds. An especially brilliant meteor might leave a recognizable trail lasting several seconds after the meteor itself has disintegrated.

Although it isn't practical to see meteors just by staring at the sky at random, there are certain times during which you will have better than average luck, and periods in which you will see meteors with less-than-average frequency. On any given night, the frequency of observed meteors will increase steadily as the hour gets later, reaching a maximum just before sunrise. This happens because of the rotation of the Earth on its axis, acting in conjunction with its revolution around the Sun.

Meteors can fall at any time, day or night, but the likelihood is greatest at a particular spot when the surface is advancing along the path of the Earth's orbit. This occurs at approximately sunrise (A of Fig. 5-15). The probability of a meteoroid entering the atmosphere is smallest at a given location when the surface is retreating along the path of the Earth's orbit; this takes place at about sunset (B of Fig. 5-15). This effect is most pronounced near the equator and it decreases with increasing latitude. At the poles, the effect is nonexistent with respect to the 24 hour rotation period of the Earth.

Superimposed on the daily cycle is a yearly cycle that results from the tilt of the Earth's axis. At the solstices—about the 21st of June and December—the Earth orbits the Sun in such a way that our planet's motion is exactly perpendicular to its axis. At those times, both poles will be exposed to an equal number of spaceborne particles (A of Fig. 5-16).

During the northern summer and autumn, the North Pole is tilted in the direction of the Earth's path around the Sun. This effect is only slight at first, just after the solstice near the end of June, but increases until it reaches a maximum at the autumnal equinox near the end of September. The North Pole is then exposed to more meteoroids than the South Pole (B of Fig. 5-16). The effect diminishes following the equinox and disappears

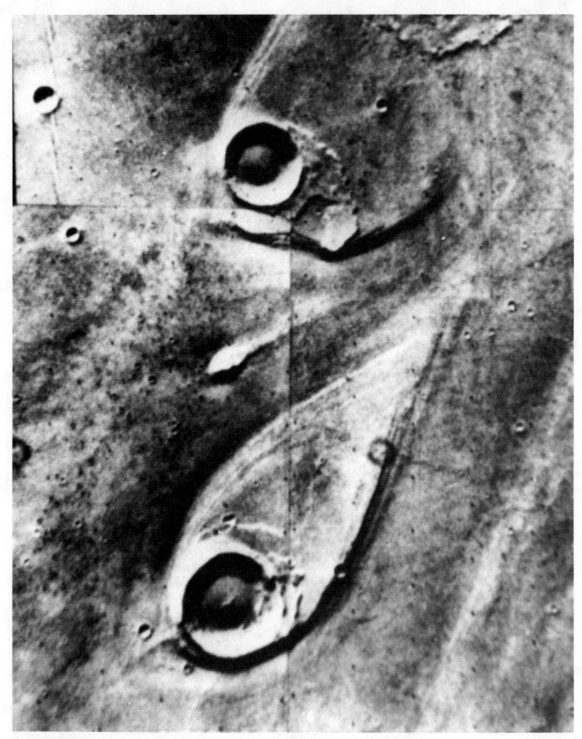
Fig. 5-14. Evidence of crater erosion on Mars (courtesy of NASA).

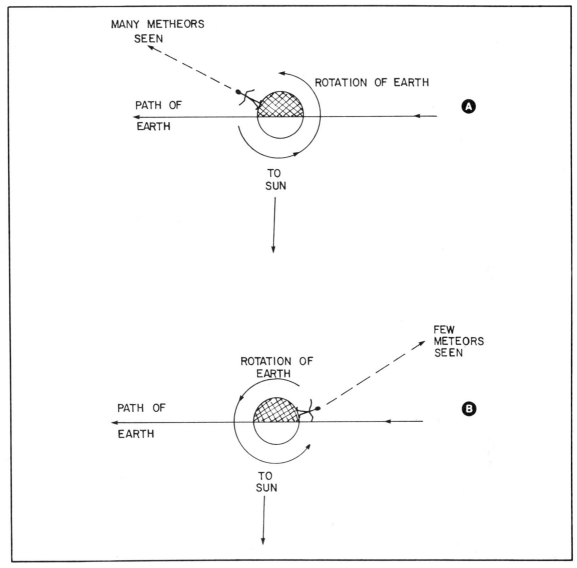

Fig. 5-15. We see the most meteors near sunrise (A) and the fewest near sunset (B) because of the motion of our planet through space.

altogether by December 21. Then the South Pole begins to turn in the direction of the Earth's path, and the tilt becoming maximum at the vernal equinox near the end of March. Then the South Pole gets exposed to more meteoroids than the North Pole (C of Fig. 5-16). The effect diminishes again after the equinox, returning to zero by the solstice on about June 21.

The yearly meteoroid cycle affects mostly the poles. It is felt less and less as the latitude becomes lower, and is practically negligible at the equator.

Overriding both the daily and yearly cycles, however, are certain brief spans of time during which the Earth passes through meteoroid swarms. These well-known showers occur at the same time every year.

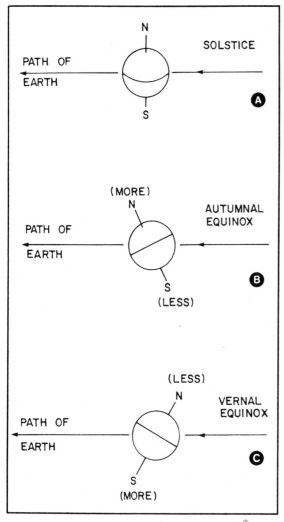

Fig. 5-16. At the solstices, either pole is exposed to the same number of meteoroids (A). Near the equinoxes, one pole or the other receives more (B and C).

Meteor showers seem to originate from particular places in the sky. Generally, showers are named according to the constellation from which the meteors seem to be coming. The name of the constellation is followed by the suffix *ids*. For example, near the end of October there is a proliferation of meteors that appear to come from Orion; this shower is known as the Orionids. Table 5-1 lists the most well-known meteor showers that take place each year.

Why do meteors seem to come from a particular spot in the sky during a shower? It seems logical to suppose that they should fall at random, coming from any direction with equal probability. This is not the case. Perspective plays a role in the appearance of meteors during a shower.

Have you ever lain down flat on your back outside during a light rain? It isn't something most people are inclined to do, but you should bundle up in a good raincoat and try it sometime. Look directly at the raindrops as they come toward you. They will seem to move outward from a certain point in the sky. If there is no wind, this point—called the radiant—will be directly overhead. If there is a light wind, the radiant will be high in the sky but not straight up. If there is a high wind, the radiant will appear low in the sky, and might be hard to identify.

Meteor showers behave exactly like the raindrops. as the Earth moves through a swarm of meteoroids, they all fall in more or less parallel paths. From the point of view of an observer on the ground, they do not seem to be parallel. Instead, they seem to come from a certain spot in the sky. This spot stays fixed with respect to the stars, moving across the sky from east to west as the night progresses. Sometimes the radiant of a meteor shower is well defined, and sometimes it is not so well defined. The meteors seem to move outward from the radiant as they fall (Fig. 5-17). The location of the radiant is determined by two things: the direction in which the Earth is moving, and the movement of the meteoroids in the swarm.

You might have noticed from Table 5-1 that a large proportion of meteor showers are named after constellations in the Zodiac (those constellations that lie near the plane of the ecliptic). This is no accident. The Earth revolves around the Sun at a fantastic speed—more than nine miles per second. This motion is always in the ecliptic plane. Therefore, we should expect that many meteors will be seen coming from the general direction of the ecliptic in the sky. Fewer meteors arrive from above or below the ecliptic plane. When they do, their relative speeds are generally less than the speeds of meteors in the ecliptic.

Meteor showers are almost always associated

Table 5-1.

Name of Shower	Approximate Dates	Associated Comet
Andromedids	Mid-November	Biela
Delta Aquarids	End of July	?
Draconids	Mid-October	Giacobini-Zinner
Eta Aquarids	Early May	Halley
Geminids	Mid-December	?
Leonids	Mid-November	Temple
Lyrids	Mid-April	1861 I
Orionids	Mid-October	Halley
Perseids	Mid-August	1862 III
Taurids	Early November	Encke
Ursids	Mid-December	Tuttle

with comets. The Beta Taurids, for example, occur as a result of debris cast off by the Comet Encke, a small, dim comet with a nearly circular orbit. The Eta Aquarids and the Orionids are thought to be associated with Halley's Comet. The andromedids are fragments of Comet Biela, which split up in the mid-nineteenth century. Comets slowly degenerate as their ices evaporate. This leaves swarms of pebbles that gradually spread out along the original orbits. It the Earth passes through the swarm, the fragments fall as meteors, and we see a meteor shower.

The most spectacular meteor showers take place when the Earth's orbit precisely intersects the orbit of the meteoroids (Fig. 5-18). This does not always happen. Actually it almost never does. The orbits might come very close to each other but seldom do they exactly intersect. Therefore, the intensity of a meteor shower changes from year to year, and once in a while we get a really spectacular show.

Meteor showers are not hard to photograph. It is only necessary to point the camera at some point near the radiant and open the shutter for a few minutes. A telescope is not needed. A wide-angle lens can be used to get a view of a large part of the sky. In the photograph, the meteors will appear as straight lines centered at the radiant. If a clock drive is used so that the camera follows the stars, the radiant will appear more well-defined than if the camera is kept stationary. If the meteor shower is especially intense, remarkable photographs can be obtained even by an amateur. It is important, of course, that conditions be optimized for astrophotography (as described in Chapter 4).

Meteors do not always fall at the same speed. The orientation of the orbit of a meteoroid, in relation to the orbit of the Earth, affects the speed at which the object will fall. Also, the speed of the meteoroid in its orbit has an effect. Meteors fall the fastest when they are in retrograde orbits lying in or near the plane of the ecliptic (A of Fig. 5-19). The slowest speeds result from meteoroid motion nearly identical with that of the Earth (B of Fig. 5-19). Meteoroids arriving from orbits well outside the ecliptic have intermediate speeds (C of Fig. 5-19).

Astronomers have various techniques for measuring the speeds of meteors as they fall. The most common method is to interrupt a time-exposure photograph at regular intervals (say a tenth or a hundredth of a second). When this is done, meteors show up as dotted lines, and the speed can be calculated from the spacing of the interruptions (Fig. 5-20). Once the speed of the meteor is known, along with its direction, the original orbit of the meteoroid can be found. This is how astronomers know that the Andromedids, for example, are associated with Biela's Comet. The orbits of the meteoroids match up almost exactly with the orbit of the original comet.

Another way to measure the speed of a mete-

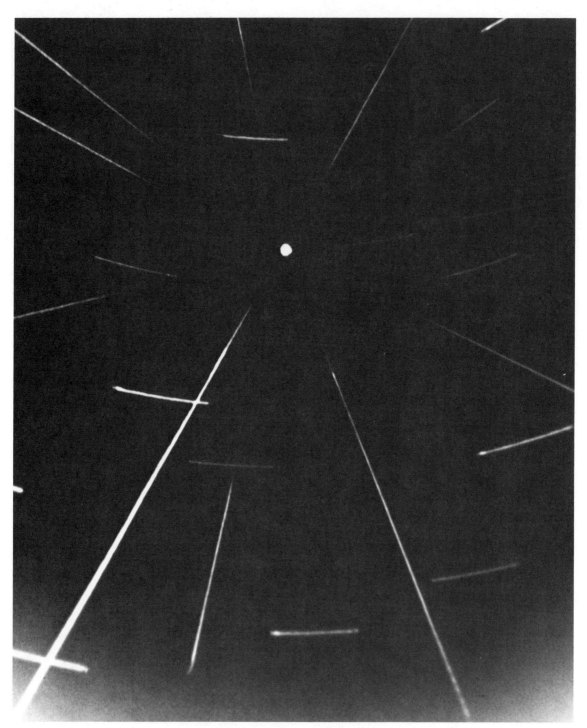

Fig. 5-17. A hypothetical time-exposure of a meteor shower. The average location of the radiant (which actually moves across the sky with the stars) is shown by the white dot.

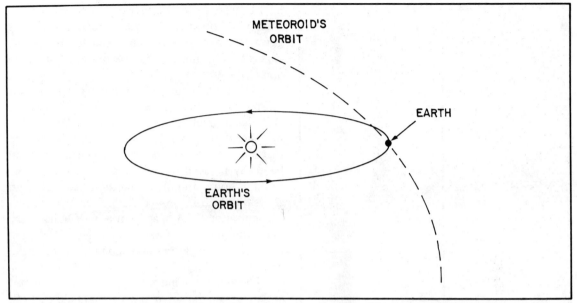

Fig. 5-18. We see a meteor shower in its most vivid form when the orbit of the meteoroids exactly intersects the orbit of the Earth.

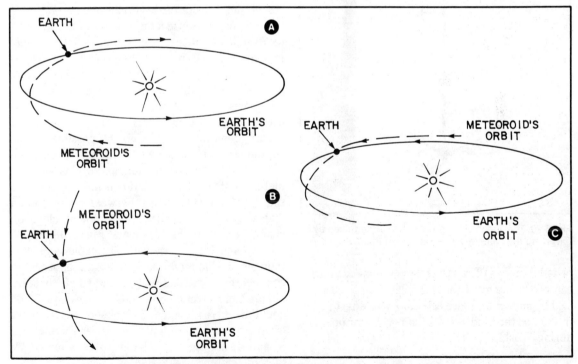

Fig. 5-19. Meteors have the highest speed when the Earth passes through a swarm of meteoroids moving in the opposite direction around the Sun (A). The slowest speeds are observed when the meteoroids move along with the Earth (B). Intermediate speeds are seen when the meteoroid orbits are well outside the plane of the ecliptic (C).

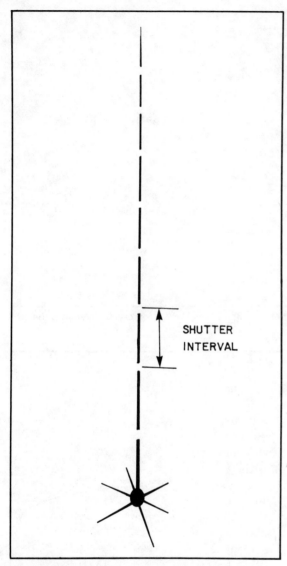

Fig. 5-20. The speed of a meteor can be ascertained by interrupting the camera shutter at known intervals.

or is by means of radar. A heavy meteor shower can produce a vivid radar display that is chacterized by numerous "false echoes." This effect raised the hair on the head of more than one radar operator during World War II!

IONIZED TRAILS

Avid amateur astronomers look forward to the annual occurrence of the various meteor showers, hoping to witness a spectacular display and perhaps get a few good photographs. Among others who eagerly await meteor showers are amateur radio operators.

When a meteor falls through the upper atmosphere, the heat produces an ionized trial. At altitudes from about 40 miles to 200 miles, the atmospheric molecules are so sparsely distributed that many of them carry an electric charge or can be easily induced to carry a charge. During the daylight hours, the ultraviolet light from the Sun strips the electrons from the atomic nuclei, producing ionization. The heat from a meteor has very much the same effect except that the ionization is more dense and occurs only along the path of the meteor.

The amount of ionization produced by a meteor depends on the size of the meteor and the length of its path through the atmosphere. Generally, the ionization persists for just a few seconds, and sometimes it lasts for less than one second.

The most intense ionization occurs at about 60 or 70 miles up in a part of the upper atmosphere known as the E region. When a meteor passes through this layer, the ionized trail will reflect radio signals. This briefly makes communication possible between two stations separated by tens or hundreds of miles (Fig. 5-21). If a meteor shower occurs, communication might be possible almost continuously. If the shower is intense, a conversation can be carried on uninterrupted. This so-called meteor-scatter communication is most common at the very-high frequencies, between about 30 and 300 megahertz. Because the common television channels are in this frequency range, television reception (using an antenna, not a cable system) is influenced. You might be watching an unoccupied channel (for some odd reason) and suddenly see a picture from a station hundreds of miles away. The next time there is a major meteor shower you might try this as an experiment. The lower channels (2, 3, and 4) are the most likely to be affected by meteor-scatter propagation.

Although meteor-scatter communication is interesting, it will probably never be relied upon. It

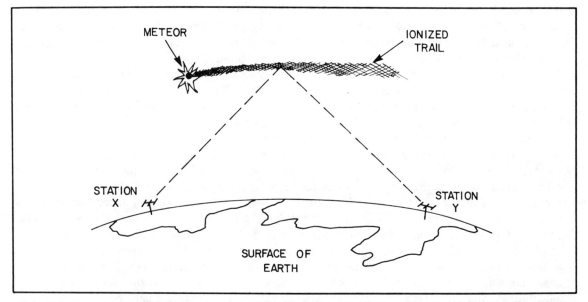

Fig. 5-21. Meteors leave ionized trails in the upper atmosphere. This sometimes makes over-the-horizon communication possible between radio stations not normally within range of each other.

is possible only during a heavy meteor shower. This is why it is of interest primarily to amateur (rather than commercial) radio operators.

WHAT MAKES UP A METEOROID?

There is only one way to tell what meteoroids are made of, and that is to find one and examine it. This has been done with meteorites after they have reached the Earth. After they are cool, fragments are brought into the laboratory and analyzed.

The two main kinds of meteorites that have been identified are stony and metallic. These are usually called aerolites and siderites, respectively. We can add a third category of meteorites: the tektites. I will put off a discussion of the tektites for a little while because they evidently are of different origin than the aerolites and siderites.

The stony type of meteoroid is very similar to a stone of Earthly origin. They are made up largely of silicate material that ranges in size from tiny pebbles to huge boulders. The metallic meteorites are made of iron and nickel. Many meteoroids contain both silicates and metal. When the objects enter the atmosphere, the heat causes them to melt; stony meteorites often have a glassy appearance on the outside. Metallic meteorites are blackened on the surface and silver-gray inside.

It is logical for us to wonder why we see meteoroids at all, and not just gases, in interplanetary space. Where did all that metal and rock come from? How did the matter condense into solid form?

To answer this question, we must look back billions of years, to a time long before the Sun shone. The whole configuration of the Milky Way Galaxy was different then. Many of the stars we see today did not exist, but other stars were there. Within these stars, hydrogen was converted to heavier elements by the fusion process. Helium was formed from hydrogen, and later more complex atoms arose. Among these elements were the silicates and metals that we find in meteoroids.

Many of the larger stars eventually blew up, scattering their matter into interstellar space. Even now, we occasionally see a star explode in our Galaxy or in other galaxies. Starstuff is scattered all over the Milky Way in the form of gas and dust and meteoroids and comets. Every atom in every meteoroid was once a part of some primeval star that exploded and threw debris into space. Our Solar System, including the Sun, congealed not only from virgin hydrogen, but also from the material that was once in other stars.

We might think of two different origins for meteoroids. We have already seen that comets disintegrate eventually, leaving meteoroids as the only entrail. Such meteoroids can be called "cometary." After the lighter, icy elements have been vaporized by repeated passages near the Sun, only the rocky stuff is left.

There is another way that meteoroids can be formed: from asteroids. Most of the asteroids in the Solar System lie between the orbits of Mars and Jupiter, but there are some that wander across the orbit of our planet, and some far out past Jupiter. Most asteroids and meteoroids orbit our parent star in nearly circular orbits, but a few have decidedly elliptical paths. Now and then a tiny asteroid—not really big enough to be called an asteroid—encounters the Earth. As a result, we see a meteor or meteorite as it heats up by friction in our sea of air and perhaps reaches the surface before disintegrating.

METEORIC DUST

Some meteoroids are so tiny that they resemble particles of sand or dust. Countless trillions of these particles are scattered throughout the plane of the Solar System. The Earth is continually beseiged by them, but they have no visible effect. They are too small to cause even the faintest noticeable flicker as seen from the ground, and they burn up even before they penetrate as far as the troposphere.

Meteoric dust does reach the surfaces of the smaller planetary moons because they have no atmospheres. We can expect, when we travel to such satellites, to find their surfaces covered with dust. Our Moon has several inches of dust on its surface. This dust is so fine that the footprints of the Apollo astronauts were clearly outlined (Fig. 5-22).

We might expect that the moons of other planets would have more or less of this dust, depending on their proximity to the Asteroid Belt. The moons of Jupiter lacking atmospheres probably have more dust than our Moon because Jupiter is nearer the asteroid belt, where innumerable collisions grind the fragments ever finer. Of course, we will not know this for sure until we land a space probe on one of these satellites.

While we cannot directly notice meteoric dust from our vantage point here on the surface of the Earth, it is possible to indirectly detect it. Every single particle of material in interplanetary space, no matter how small, reflects the light of the Sun. Whether it is the size of a house, a baseball, a pea or a grain of sand, each meteoroid has phases similar to those of the Moon.

When a particle is nearly in between us and the Sun, its phase is crescent and we see very little reflected light (A of Fig. 5-23). When a particle is at an elongation of 90 degrees (B of Fig. 5-23), it looks brighter because it is at half phase. When a particle is nearly at opposition, its phase is full and it reflects the most light back toward us (C of Fig. 5-23).

The particles are too small for us to be able to see the phases, even through the most powerful telescope, but the phases nevertheless exist, and their effect on particle brightness is just as great as the effect of the phases of the Moon. When a meteoroid is exactly opposite the Sun, all shadows disappear as far as we can tell, and the whole surface is sunlit. The brightness at full phase is much greater than at any other phase (even three-quarters or nine-tenths).

On a particularly dark night, when the sky is clear and there are few man-made lights nearby to interfere with viewing, the ecliptic can be seen as a faint, hazy band arching across the sky. This band, called the zodiacal light, is most easily seen on spring evenings, after twilight has faded, or on autumn mornings before the first light of the coming dawn. The zodiacal light can also be seen during the wintertime around midnight. The hazy band is caused by light reflected from the meteoric dust in the plane of the Solar System.

The zodiacal glow can easily be confused with the Milky Way Galaxy, but the difference is evident because the Milky Way does not lie along the ecliptic. At the above-mentioned optimal viewing times, the zodiacal light band runs from east to west

Fig. 5-22. Meteoric dust on the Moon (courtesy of NASA).

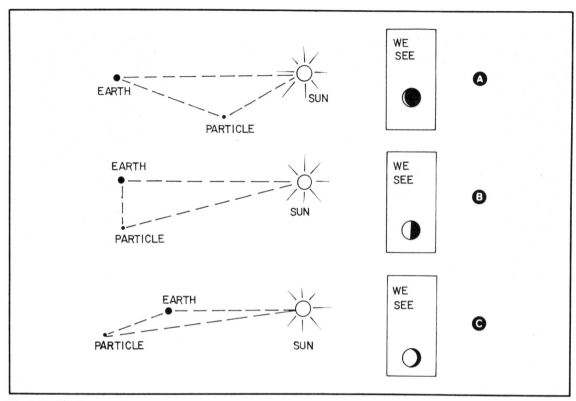

Fig. 5-23. Small meteoroids in space reflect varying amounts of light toward the Earth, depending on their positions relative to the Earth and Sun. At A, little reflected light is seen, at B, an intermediate amount is seen and at C, a large amount is seen.

and almost directly overhead at the temperate northern latitudes.

Because of the effects of phase, the zodiacal light appears brightest near the point opposite the Sun (from where we receive the most reflected light). On spring evenings, this point is near the eastern horizon. In the predawn hours during the fall, this point is low in the western sky. In the middle of a winter night, the brightest part of the zodiacal glow band is almost straight overhead, and it is then that it is the most easily seen. It appears as a diffuse, fuzzy, whitish spot.

Astronomers call this the counterglow, and often used the German word gegenschein. Counterglow does not occur because there is a greater concentration of particles opposite the Sun than elsewhere, but only because of the effects of phase. Particles near the counterpoint (the direction in the sky opposite the Sun) are all in full phase and look much brighter than particles in other parts of the heavens.

You might have noticed counterglow in Earthly situations. If you have flown over clouds in an airplane and looked for the shadow of the plane on the clouds, you have probably noticed the counterglow of sunlight reflecting off of the water droplets. You can stand in your yard on a sunny afternoon and notice the counterglow surrounding the shadow of your head. These effects occur for exactly the same reason as the celestial gegenschein. At the counterpoint, all shadows vanish, and the optimum amount of light is reflected toward you. This is true almost regardless of the nature of the substance at which you are looking.

LAGRANGIAN POINTS

Although meteoroids do not concentrate themselves directly opposite the Sun with respect to the Earth, there is some evidence to suggest that particles might have a propensity for gathering in certain places because of gravitational effects. The idea was first suggested by the mathematician G. Lagrange in the eighteen century. Lagrange proved that objects at the verticles of two identical equilateral triangles, inscribed in a certain orientation within the orbit of those objects around a larger mass, have the property of maintaining their positions.

Specifically, if a planet orbits the Sun at a certain distance, call it r (for radius), and there are other, smaller masses orbiting the Sun that happen to come near the planet's orbit at a distance r from the planet, those other masses will have a tendency to remain near those points. This effect is diagrammed in Fig. 5-24.

The special points, which astronomers call Lagrangian or libration points, precede and lag the planet along its orbit by 60 degrees of arc. The two triangles determined by the Sun, the planet, and the libration points are equilateral; that is, their sides are all of length r.

Lagrange proved that gravitational effects would tend to favor the accumulation of matter at the libration points. This might happen for any of the planets, including the Earth. It might also happen in planet-moon systems, especially in the case of Earth and its relatively large natural satellite.

It was almost a century and a half after Lagrange's proof before anyone found an example of this effect operating within the Solar System. The search had been carried on, unsuccessfully, with most of the attention given to Jupiter, because it is the most massive planet. With the improvement in observing apparatus and techniques, numerous small asteroids were found concentrated near the libration points in Jupiter's orbit around the Sun. The asteroids persistently moved in paths that led and lagged Jupiter by 60 degrees of arc—at a distance equal to the radius of the massive planet's orbit—or a little less than a half billion miles.

It is likely that there are countless smaller particles near these points, too small to be seen by our telescopes, preceding and following Jupiter like schools of fish. Similar swarms probably also exist in the orbits of the other large outer planets (Saturn, Uranus, and Neptune). We would expect these

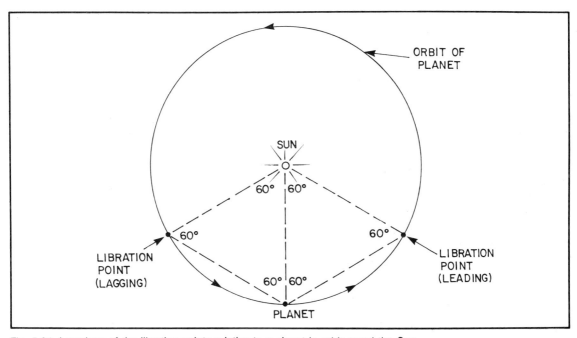

Fig. 5-24. Locatinos of the libration points relative to a planet in orbit round the Sun.

swarms to be smaller than those associated with Jupiter, however, because of the greater gravitational pull of the Jovian giant, and also because of the greater concentration of asteroids in that region of the Solar System.

Are there meteoroid/asteroid clusters near the libration points in the Earth's orbit? Probably, and the other inner planets (Mercury, Venus, and Mars) most likely have them, too. It would be wise for space travelers to avoid the libration points in any planet's orbit—especially in the case of Jupiter. The probability of a collision with a meteoroid or asteroid would be much larger in such regions than in other parts of the Solar System (with the possible exception of the Asteroid Belt itself).

Once the swarms were found preceding and following Jupiter, astronomers began to wonder whether they would also be able to find them leading and lagging the Moon in its orbit around the Earth. If such swarms did exist, astronomers reasoned, they would probably not be easy to find. If it were orbiting our planet at a distance of only a quarter of a million miles, any major asteroid would certainly have been discovered long ago. Consisting of tiny meteoroids, the swarms would probably be extremely faint and diffuse.

Photographic methods were employed—as well as direct observation via telescope, binoculars, and the naked eye—to find the expected clouds. The task was complicated and there were disagreements among various scholars as to whether or not the Earth-Moon libration swarms actually had been seen. The possibility that the Earth might have many natural satellites (besides the Moon) was intriguing. Proof of their existence came hard; the controversy still rages.

In spite of the lack of direct physical evidence for meteoroid swarms leading and lagging our Moon, if I were the captain of an interplanetary vessel, I would steer clear of the two theoretical libration points as given by Lagrange's theorem!

THE MYSTERY OF TEKTITES

Strange stonelike objects have been found in various places on the Earth. The odd, glassy stones were named tektites by the scientist and geologist F. Suess at the beginning of the twentieth century. Most scientists agreed that they had to be objects from space. The objects resemble rocks of volcanic origin. Despite their wide distribution on our planet, they are similar in composition. These objects appear to be fused as if they have encountered extreme temperatures. The silicate matter has been melted to form a glasslike substance. Their rounded shape also indicates that they have probably been heated. It is as if they were cast down from the sky onto terrain that bears practically no resemblance to them.

These puzzling objects have been found all around the world, but they seem to be concentrated in the lower latitudes (especially in Australia and southeast Asia). They have been discovered in lesser quantities in southern North America and southern Europe. A few have been found in Africa.

Tektites were found in places totally devoid of volcanoes so it was doubtful they were made of cooled-off lava from volcanic eruptions (unless some monstrous volcano once blew up with such violence that the lava was thrown nearly into orbit).

When geologists measured the ages of the tektites using radioactivity methods, they got a surprise. Some of the objects were less than a million years old. If scientists could believe their findings, the tektites could not have been formed along with the Solar System (which was born billions of years ago). Could there have been some kind of cosmic catastrophe that made the tektites?

With the consensus that tektites must have come from space, certain questions had to be answered. Why were they so much different from the rocky and metallic meteorites? Why were they so young, comparatively, with respect to other kinds of meteorites? Why were they concentrated only in certain parts of the world?

Various theories were presented to explain the origin of the tektites. Different tektite "groups"—that is, tektites from different concentrated areas—differ in age, although all of them in a given group are about the same age. Perhaps there were several rather than just one catastrophe. One event would be responsible for the tektites in Australia and Southeast Asia, another would have produced the tektites in North America, and still another event would be responsible for the objects found in Europe and Africa. What sorts of events would these be?

Most astronomers of today believe that the tektites rained down from space as a result of a major meteorite impact on the Moon. A massive meteorite, perhaps more aptly called a small asteroid, could precipitate an explosion on our Moon of such proportions that some of the debris might be cast out of the Moon's gravitational influence. The moon is much less massive than the Earth, and its gravitation is far less intense. A small asteroid, weighing billions of tons and moving at several miles per second, could easily blast moonstuff into interplanetary space. What if some of this material fell onto the Earth? That would explain the tektites, and in a pretty dramatic way.

Those of us who have looked at the Moon through even a small telescope know that our satellite planet has been bombarded by large meteorites. The huge craters bear witness to that. Some of the large, flat "seas" or valleys are the basins for the oldest craters formed when the Solar System was very young and littered with many errant asteroids. Some of the collisions were probably so cataclysmic that the Moon was nearly shattered. Surely the earth was assaulted by fiery Moon rocks billions of years ago. We would not be likely to find any of the tektites from those earliest impacts, but younger ones—created by the impact of a small asteroid within the past few million years—might still be found.

The Moon has many craters with prominent radial lines called rays. It is not difficult to understand the origin of the rays if we think for a moment about the violence of an asteroid impact. Matter from the Moon was thrown outwards, splashing away in all directions from "ground zero." Some material was no doubt thrown high into the sky, most of it falling back to the surface, but some escaping into space. The crater Tycho has the most prominent rays of any Moon crater.

Moon rocks were examined by unmanned landing vessels during the 1960s. The composition of the rocks in the crater Tycho greatly resemble that of the youngest tektites. Tycho seems to be a fairly new crater on the Moon; that is the main reason why its rays are so prominent. It has been found that some of the Moon rocks ejected from Tycho during the impact—somewhat less than a million years ago—could have reached the Earth.

Suppose that our astronomers and geologists are correct, and the crater Tycho was formed less than a million years ago. Manlike creatures roamed the Earth then, and they probably had beliefs about the Sun and the Moon and the stars. They must surely have noticed how the Sun's course across the sky changes with the seasons. Certainly they saw how the Moon goes through phases. The heavens, to them, must have represented stability and regularity as well as vastness. Imagine what the ancestors of man would have thought if they saw the fiery explosion as a small asteroid plunged into the Moon. The stability of the universe would be ruined. The people would have been terribly afraid.

There is a good chance that our forebears did witness the birth of the crater Tycho. Thousands of years later, a hailstorm of fiery, glasslike stones, one of countless swarms of moonstuff thrown into space by the impact, took place on the Earth in the part of the world we now call Australia, Vietnam, Cambodia, Laos, Thailand, and other countries of the Western Pacific. This "group" of tektites is the largest yet found. Other smaller, older "groups" probably resulted from earlier asteroid strikes on the Moon. Whether those collisions were any less cataclysmic than the one that formed the Australian/Asian tektites is open to question. Time erases everything, and tektites can be no exception. The African and North American objects, while seemingly fewer in number, might have largely disappeared because they are much older.

THE ASTEROIDS

In Chapter 1, there is a description of how the mathematician Bode found a correlation between certain numbers and the locations of planetary orbits in the Solar System. For the Bode number 2.8, there was a mysterious gap: No planet existed at a distance of 2.8 astronomical units from the Sun.

While it is difficult to imagine how Bode's sequence could bear any more than a coincidental relationship to the masses orbiting the Sun, some astronomers took it seriously enough to spend a lot of time looking for the missing planet.

In a sense, they did find it, but it did not exist as a discrete mass. Instead it is innumerable fragments of debris moving slowly around the Sun in nearly circular orbits.

Italian monk and astronomer Giuseppi Piazzi found the first asteroid on New Year's Day in 1801 (right at the beginning of the nineteenth century). The search for the missing planet was over, but the results were somewhat of a disappointment. The new planet was in almost exactly the correct location, but it was so tiny that it hardly deserved to be called a planet. The object was named Ceres, after the Roman goddess of agriculture. When the length of Ceres was determined, it was found to be only about 400 miles long.

It was not long before other asteroids, even smaller than Ceres, were found at a distance of about 2.8 astronomical units from the Sun. Eventually it was determined that there are millions of asteroids and most of them orbit in several nearly circular bands. They are something like the rings of Saturn, but on a vastly larger scale. The inner asteroids move faster than the outer ones. There is a nearly two-to-one spread in the orbital periods of the asteroids in the main belt between Mars and Jupiter. Some revolve in about three years; others take about six years.

The asteroids do not all follow perfectly circular orbits, and they do not all orbit in exactly the same plane. In this respect they are like the planets. Some asteroids have more elongated orbits than others, and orbits more or less tilted with respect to the plane of the Solar System. The result is not perfect order; it is rather like a huge swarm of bees all moving in generally the same manner, but each individual following an individual path. The result is continual collisions that make for an unending series of minor cosmic catastrophes. Pieces break off of asteroids when they run into each other, and this forms smaller and smaller particles. Given enough billions or trillions of years, the Asteroid Belt will grind itself into a ring of sand and dust.

Many asteroids are known to be irregular in shape, and to tumble (rotate) on an axis. This we would expect because of the innumerable collisions. We know this because some asteroids change brightness periodically, reflecting more or less sunlight depending on which side is facing us. For example, we might see a generally rod-shaped asteroid first broadside and then an end-on view (Fig. 5-25). It would look brighter first and then dimmer.

While the majority of asteroids lie in the belt between Mars and Jupiter, there are some that do not. One example is the pair of swarms that precede and follow Jupiter by 60 degrees of arc. A few asteroids are known to follow orbits inside that of the Earth, and some asteroids periodically cross the orbit of our planet.

Several asteroids are known to periodically come quite close to the Earth. Among these, Adonis and Apollo have perihelion distances of less than one astronomical unit. Their orbits are quite elongated; the aphelia are about as far away from the Sun as the main Asteroid Belt. Eros, an asteroid 15 miles in diameter with an almost circular orbit lying entirely outside the orbit of the Earth, has come within just a few million miles.

Adonis and Apollo are of special interest because they are occasionally at exactly one astronomical unit from the Sun. That is the same distance from the Sun as the Earth. Does this mean that either of these two objects will ever strike our

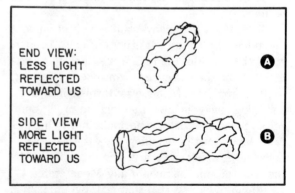

Fig. 5-25. The brightness of an irregular meteoroid is affected by its orientation relative to our line of sight. At A, the hypothetical meteoroid appears much dimmer than at B.

planet? Not as long as their orbits maintain their present tilt relative to the ecliptic. When either of these asteroids are at the same distance as the Earth from the Sun, they lie either above or below the plane of our planet's orbit. The orbit of the Earth does not intersect the orbit of either Adonis or Apollo (at least not as long as nothing changes). It is possible that gravitational interaction with the Earth could eventually modify the orbit of Adonis or Apollo in such a way that a collision could take place.

If a large, known asteroid such as Adonis or Apollo were to fall onto the Earth, the resulting catastrophe would probably exterminate humanity. If the object came down on a land mass, the whole planet would be shaken by earthquakes, gigantic tsunamis would circle the globe, and volcanoes would erupt, their smoke spreading throughout the atmosphere and blocking out the Sun. If the object landed in an ocean, the disaster would be similar, but the tsunamis even higher. We can take comfort in knowing that, with today's technology, we would be able to detect the asteroid as it approached. Perhaps we could send space probes, equipped with huge nuclear warheads, out to change its orbit enough to prevent the doomsday.

Asteroids are still being discovered. It is doubtful that we shall ever know all of these pieces of rock having diameters of less than a mile or so. Certainly there are countless asteroids big enough to fill an auditorium, but too small to be detected from our vantage point on Earth (unless they come extremely close).

It is the smaller asteroids that should worry us more than the larger ones. A small asteroid is more difficult to find. We might not notice it until there was nothing we could do to stop it from hitting us. And there are a lot more small asteroids out there than large ones. While the impact of an asteroid a hundred feet across might not kill off humankind, it would certainly cause dramatic consequences for much of the world.

It is not my intention here to arouse mortal fear of malevolent stones assaulting the Earth next week, next month, next year, or even in the next hundred years. The chances of a major impact occurring within your lifetime are practically zero. Nevertheless, even the most improbable things eventually happen, given enough time. There is a fair chance that humankind will see a small asteroid come down within the next 100 centuries. Increase the allotted time to a million years and the chances become better. Give it a billion years and the improbable turns into the virtually inevitable.

Let's consider a numerical example. Suppose that the chances of an asteroid at least a hundred feet across landing within the next year on a populated area are one in a million. That means that we have an 0.999999 chance of escaping. In a period of two years, the probability of escaping is 0.999999 times itself, or about 0.999998. In three years, this goes down to 0.999999 cubed, which is about 0.999997. That's about a three-in-a-million chance of getting hit. Still not much. After a century, the chance of our staying safe is 0.999999 to the hundredth power (or approximately 0.9999). We still have only a one-in-ten-thousand chance of disaster.

Consider a period of two thousand years. That's the length of time since the zenith of the Roman Empire. The chance of staying safe is now 0.999999 to the two thousandth power (or about 0.998). Probability of catastrophe is two in a thousand. The chances are getting slightly, but inexorably, bigger. Have you ever won a large lottery prize?

As the time frame is increased, the chance of a strike continues to get larger (although for very long times the above mathematical model gets less and less accurate). Besides, the original estimate of a one-in-a-million chance in the next year is probably way off. As this example serves to show, time overcomes all obstacles.

FINDING AND TRACKING ASTEROIDS

Asteroids are located in very much the same way as the planet Pluto was discovered in 1930. A large telescope is necessary in order to gather enough light so that extremely faint objects can be seen. The brightest asteroids have long since been catalogued; asteroid hunting is now an art more or less reserved for the professionals who have access to large observatories.

An asteroid might be found in two ways. Both methods involve photographing a particular region of the sky. Asteroids move with respect to the background of stars (even though they appear as points of light, just as stars do). Therefore, in a lengthy time-exposure photograph using a clock drive to keep the telescope pointed at the same part of the sky, an asteroid can be recognized as a short line while the stars appear as points. Another method of finding asteroids is to take two photographs, several hours or days apart, and then compare them.

Looking for miniscule differences in two films, each containing thousands of stars, requires patience. A special instrument can be used to simplify the process. The two photographs, usually negative transparencies, are placed in a machine that allows viewing of either one alternately through a magnifying scope. Switching between one negative and the other presents the exact same picture of the stars; the negatives are lined up with extreme care. Anything that has moved relative to the background of stars will then seem to jump back and forth as the viewing scope is switched from one photograph to the other and back again. With meticulous examination of the two plates, asteroids are found from time to time.

When an asteroid is found, there are two things that must be ascertained about it before it is called "new" and given a name. First, it is necessary to be sure that it really is an asteroid an not some other kind of celestial object such as a comet. Second, the astronomer must be sure that, if the object is an asteroid, it is really "new" and not already catalogued.

It might not be immediately apparent whether an object is an asteroid or a comet. Most asteroids orbit quite close to the ecliptic, and in the same direction as all of the planets. If an unknown object is found orbiting counterclockwise (as seen from the north pole of the Solar System) and very nearly in the plane of the ecliptic, we have good reason to think that it is probably an asteroid. Comets can follow orbits inclined at any angle. They have retrograde motion around the Sun just as often as they have forward or "normal" motion.

Nevertheless, the fact that an object is in the ecliptic, and has a forward orbit, does not mean that it cannot be a comet. Such coincidences can and do occur. It takes some time to ascertain the nature of a newly found celestial wanderer.

Once an object has been determined to be a "new" asteroid, it must be catalogued according to its orbit. This requires several observations so that its path through the heavens can be accurately plotted. Then it must be named. There are only so many Greek and Roman names to go around, and we have long since run out of them. Therefore, asteroids (like comets) are now named according to the year in which they are found, and in serial order by year.

One of the major organizations devoted to the search for new asteroids is the NASA/Planetary Society Asteroid Project. It is managed by the World Space Foundation. They are committed to the search for life on other planets and in other solar systems, as well as to space research in general. Time-exposure methods are used to look for previously undiscovered asteroids.

The constant vigil often bears fruit. Recently a new asteroid was discovered by E. F. Helin and R.S. Dunbar of the NASA Jet Propulsion Laboratory, in conjunction with A. Barucci of the European Space Agency and S. Swanson of the California Institute of Technology. The new asteroid, known as an "Aten" asteroid because its mean distance from the Sun is less than one astronomical unit, was found at the Mount Palomar Observatory using the Schmidt telescope. At the time of discovery, there were three other so-called "Aten" asteroids known.

The search for "new" asteroids will never end. Certainly there are far more small asteroids we have not yet found compared with the number of relatively large ones we have catalogued. We can be sure, however, that the search will get more and more difficult and require increasingly meticulous observations.

HAZARDS FOR SPACE TRAVELERS

It is not hard to imagine what an asteroid might do to a fast-traveling spacecraft. More than one science-fiction story has depicted the consequences of a collision between a spacecraft and a massive piece of rock at a relative speed of several tens or hundreds of miles per second. What is the real danger presented by asteroids to future space-faring people?

Obviously, the likelihood of a collision with an asteroid is greater in some parts of the Solar System than in other parts. Space travelers would avoid regions of the greatest risk. For example, we would steer clear of the libration points of large planets such as Jupiter, Saturn, Uranus and Neptune. Even the smaller planets probably have swarms of small asteroids and meteoroids preceding and following them by 60 degrees in their orbits.

The biggest problem for interplanetary travelers will be the main asteroid belt between Mars and Jupiter. It will be practically impossible to avoid navigating through this region if we are ever to visit the outer planets. In order to get to Jupiter, Uranus, Neptune, and Pluto without passing through the main Asteroid Belt, it would be necessary to leave the plane of the Solar System. This would require immensely more fuel than a course within the plane.

Space probes have negotiated the Asteroid Belt and come through undamaged. The danger can be overestimated. Traveling among planets is done on a "catch-up" basis, moving in an ever-widening spiral at an orbital speed very close to that of other objects at the same distance from the Sun. While passing through the Asteroid Belt, the speed of a spacecraft relative to the debris would be comparatively small, representing only the outward-bound component of motion (Fig. 5-26).

With sophisticated instruments on board a spacecraft, future astronauts should be able to avoid collisions with asteroids moving in the general

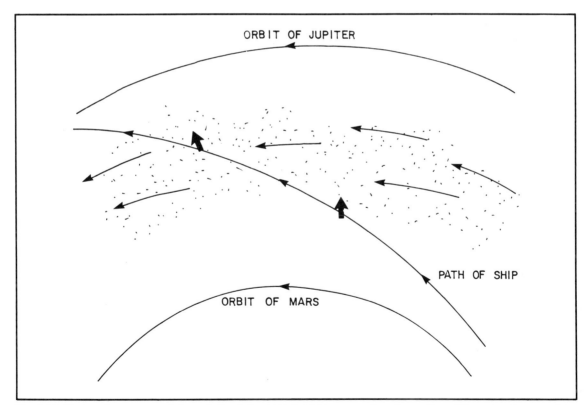

Fig. 5-26. Motion of a space craft through the Asteroid Belt. Light arrows show the motion of the asteroids around the Sun. Heavy, short arrows indicate the motion of the ship relative to asteroids in its immediate vicinity.

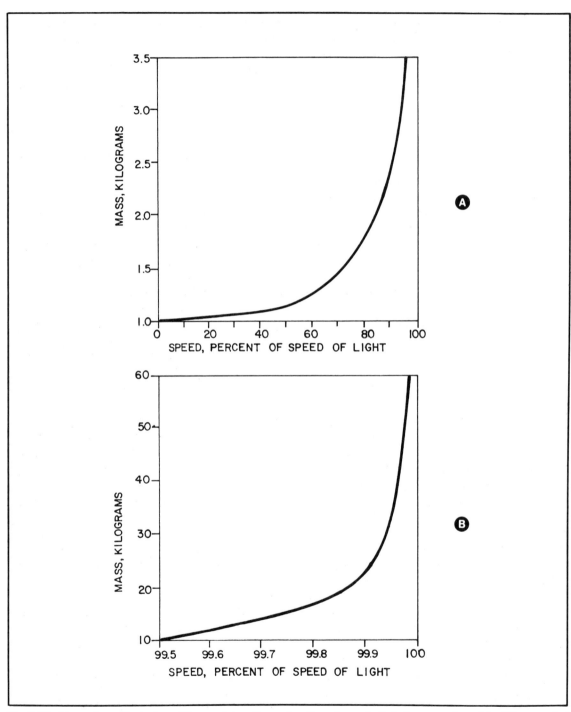

Fig. 5-27. Mass of an object of one kilogram (at rest) as a function of speed. At A, the range from 0 to 100 percent of the speed of light. At B, the range from 99.5 to 100 percent of the speed of light. There is theoretically no limit to how much the mass can increase.

direction of the swarm in the vicinity of the ship. If a collision with a smaller object does occur, it will not have very much force, and perhaps protective shields can eliminate the chances of damage.

The more worrisome problem is the errant asteroid or large meteoroid that might approach with such speed that it could not be detected before a mighty impact took place. A chunk of rock the size of a basketball, moving at a hundred miles per second, would surely cause disaster if it hit a spacecraft made of any ordinary material!

Space debris must be taken into account when we plan trips to other planets or even to the Moon. For interstellar voyagers—should humankind ever develop the technology needed to reach other solar systems—the danger will be no less great. The possible consequences will be much worse because the relative speeds will be greater. Even though their concentration is probably much smaller than within the confines of the Solar System, asteroids, meteoroids, and comets float around in all directions in between the stars.

Spacecraft will have to attain fantastic speeds—many thousands of miles per second—if we are ever to reach the stars. At a speed of a hundred thousand miles per second, even a small pebble could conceivably be propelled right through the hull of a spacecraft and out the back end with almost no difficulty. The faster we go, the more likely it will become that we will hit a piece of interstellar debris within a certain amount of time.

As if that isn't bad enough, extreme speeds produce another effect. That is the relativistic increase in mass that occurs when a vessel travels at a sizable fraction of the speed of light. At speeds of less than about 100 thousand miles per second, the effect is not too pronounced, but at 90 percent of the speed of light, the apparent mass of approaching meteoroids, asteroids, or comets would be more than doubled. At 99 percent of the speed of light, approaching objects become seven times as massive. Their density (and penetrating power) increases accordingly. Suppose we encounter a rock that has a mass of 1 kilogram (a little over 2 pounds). Its mass as a function of speed is shown in Fig. 5-27.

To reach the farthest stars, we will have to attain speeds extremely close to that of the speed of light. This will cause time to be compressed and will make it possible for us to travel all the way across the Galaxy in the span of a human lifetime.* As the speed of a vessel gets very close to that of light, the mass increases without limit, and so does the density. A small boulder might attain the mass of the entire Earth! A spacecraft struck by such an object would surely be destroyed and its occupants instantly killed.

All of those stray pieces of matter in space certainly are to be respected by future astronauts, but they might also be useful. Asteroids are no doubt rich in many of the minerals that we find in the crust of our own planet: iron, copper, silicates, and probably also silver, gold, and other precious substances. It is not entirely ridiculous to suppose that someday we will have the means to mine these minerals at reasonable cost. This will be especially practical if we set up bases on any of the larger asteroids, such as Ceres, or if we build solar-orbiting space stations not near any of the planets.

---

*This effect is described in Chapters 4 and 5 of *Understanding Einstein's Theories of Relativity: Man's New Perspective on the Cosmos* (TAB book No. 1505).

Chapter 6

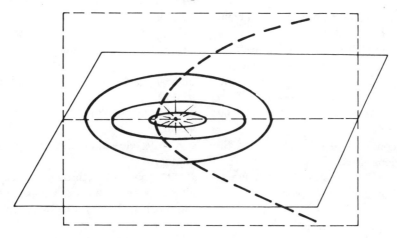

Catastrophism, Science, and Heresy

I THINK IT IS APPROPRIATE TO BEGIN THIS chapter with a parable.

A seasoned mountain climber approached the summit of what he believed to be the highest mountain in the world. No one else, to his knowledge, had ever conquered the peak before. When the tired climber finally pulled his body up over the last ledge, he noticed to his dismay that there was another person already there.

As the climber approached the person, he saw that it was a wizened old man, with an expressionless face, staring into space. The climber asked, "How did you get here?"

The old man turned to the climber and replied, "I climbed, just as you did."

"Why do you just sit here?" asked the climber.

"I have conquered the highest mountain in the world," said the old man, "and there is nothing else for me to do."

The climber felt a pang of disillusionment. No further to go; he had reached the limit. Then the climber thought to himself, "Suppose I am wrong, that this is not the highest peak in the world." How would he be able to prove that? His skepticism made him happy; maybe there was more to be explored.

Without saying anything, the climber took out a piece of apparatus designed to locate the horizon precisely in the irregular terrain of the mountains. It consisted of a telescope, equipped with a carpenter's level, so that it could be adjusted to point straight to the horizon. A pair of cross hairs, like those in a gunsight, would mark the exact tangent line to the surface of the planet at that point. If any peak, nearby or distant, rose above the point of the cross hairs, that would mean the other peak was higher than the one on which the two men rested.

The climber peered through the eyepiece and slowly rotated the telescope. Sure enough, it was not more than a few moments before the climber saw a peak, not far off, that rose above the point of the cross hairs. Checking carefully for several minutes more, the climber ascertained that no other peak in sight, except that one, rose above his present location.

All this time, the old man watched with a blank expression on his face.

The climber said, "This is not the highest peak in the world. You are mistaken. And so was I."

The old man replied, "Rubbish! I have conquered the highest mountain in the world. There is no more." And he resumed his vacant stare into space.

"Suppose I could prove it to you," said the climber. "Come look through this telescope. You know the principles of physics well enough to believe me if I show you how it works and what I have found."

The old man muttered something under his breath, but got up and came over and examined the apparatus. Then the old man looked through the eyepiece and saw the stark top of the other mountain, rising above the point of the cross hairs.

"Rubbish," the old man repeated. "This is the highest peak in the world. The ground under the other mountain is higher than the ground under this one, that is all. This mountain is really higher than that one."

The climber could not argue with that kind of statement, which seemed so patently absurd as to be ridiculous. But the climber knew he was right and the old man was wrong; he had the satisfaction of knowing that there was more to explore, that he had not yet reached the limit.

The climber packed up his equipment and went over to the other peak and climbed it—all the way to the top. Setting up the leveling scope again at this peak, he scoured the horizon and found that there were no other mountains that came up to the point of the cross hairs. One distant peak just about lined up—but no, it was a tiny bit lower.

The climber shouted to the sky, "I have found the highest mountain in the world!" Then he sat down on a rock to contemplate his find. He had gone as high as anyone could go. He began to look off into space, wondering what more he might accomplish. It depressed him a little to think he had reached the limit.

The climber lost track of time; he did not know how much longer it was before a young boy walked up to him and said, "Why are you sitting there?"

The climber, startled, looked over and said, "How did you get here?" And the boy replied, "I climbed up, just as you did."

The seasoned climber said, "I have reached the highest peak in the world. There is no more for me to do."

The boy blinked. "How do you know that?" he asked.

"My telescope tells me so," answered the climber.

The boy looked through the scope that was still aimed at the distant peak that almost came up to the point of the cross hairs. Then the boy pulled out a pencil and a pad of paper and began scrawling some numbers, letters, and other symbols on it.

"You are wrong," said the boy. "That peak, far off in the distance, at which your telescope is pointed, is higher than this one."

"Nonsense!" scowled the seasoned climber. "My instrument is perfectly calibrated. This is the highest peak."

"No," insisted the boy. "For the whole world is curved, and the horizon is not really as you see it through this telescope. You might say that the ground under the far mountain is lower than the ground here, and that the peak, while seeming lower, is actually higher!" The boy showed the climber his calculations and explained them. They were flawless.

"Nonsense!" repeated the climber, who had by now started to look aged, and resumed his vacant stare into space.

The boy, undaunted but knowing the seasoned climber would never be convinced, set off for the distant mountain. But not after obtaining the climber's telescope. The climber obviously had no further use for it.

Finally the boy reached the top of the distant mountain where he set up the telescope, making sure it was level. No other peak came anywhere near the cross hairs in the eyepiece. The boy spent days observing every mountain, calculating their heights, and taking into account the curvature of the planet. Not a single peak could be as high as the one on which he stood.

The boy shouted, "I have reached the highest mountain in the world!" He was in ecstasy; he felt like the king of the universe. But there was nobody there to tell him that, far beyond the horizon and completely out of sight, there were many mountains far bigger than his mountain—in fact, their peaks were higher than the boy could even have begun to imagine.

SCIENTIFIC METHOD

The pursuit of knowledge is a never-ending challenge. We must not get complacent and think we know everything. If we do, we are lucky if someone comes along and figuratively shakes our shoulders and wakes us up. We are unfortunate if this does not happen. In the history of science, there have been both lucky and unlucky times in this respect. Right now we live in a period of good fortune. Inquiry and skepticism are the rule, and dogma has been put largely aside—largely but not totally. Some of us insist that we have climbed the highest mountain in the universe of knowledge. The higher a scientist gets, the fewer people there are who dare to tell him that there is further to go. In the parable, first the old man, then the climber, and finally even the young boy arrived at the conclusion that they had "found it all." All were mistaken.

Scientists, including astronomers and cosmologists, operate according to certain principles of logic. There are theoreticians and experimentalists. The theoreticians try to build models of reality that will explain, accurately, what the experimentalists observe. This is an imperfect process that never stops. The more we learn the more we know there is to learn. One experimentalist keeps a dozen theorists busy. The thirteenth theorist becomes arrogant and starts to think that he is immune to inquiry.

The modern scientific method differs radically from the practice of science in times past. Hundreds or thousands of years ago, scientific facts were dictated by a privileged few and their theories carried the force of law. No one dared to question the absolute truth, lest he or she be dealt severe punishment. A notable example of this kind of "scientific method" can be found in the case of Galileo Galilei. Galileo was condemned to house arrest after proclaiming such "heresies" as the existence of mountains on the Moon and satellites in orbit around Jupiter.

Perhaps things were simpler when people took the word of Aristotle and Ptolemy at face value. These and some other ancients were practically made into gods! Nowadays, we have to work according to logic, building our theories in logical fashion, basing our knowledge on certain axioms, procedures, and experimental data. It is almost as though we must "play computer." We might even say that it reduces the human mind to the status of a programmed electronic circuit that follows, in rigorous methodical fashion, its process along the winding road of scientific inquiry. Emotion plays no part, we might think, other than to provide the drive and energy to carry on.

Logic

Logic is at the root of all scientific method. The rules of logic can be expressed in purely mathematical form, and there is not one discipline of science that uses any system of logic that differs from the commonly accepted one. We may say, for example, that A implies B (meaning: if A is true, than B is true). If your car battery is dead, your car won't start. We can change that around to say that if B is false, then A is false (if your car starts, the battery is not dead). Those two statements are logically equivalent; they convey exactly the same meaning. But we cannot say that B implies A: If your car won't start, then the battery must be dead. That might not be the case. The battery might be in perfect condition, but the gas tank might be empty.

Too often, professional as well as amateur scientists make the preceding logical mistake. That reasoning "glitch" is probably the most commonly abused, and it can result in totally wrong conclusions. If there was a Big Bang much too powerful for gravitation ever to overcome, then the universe must be expanding. According to observation, the universe is expanding, but that does not necessarily dictate, logically, that there was a Big Bang. We have to find more evidence for a Big Bang than the expansion of the universe. Astronomers of the twentieth century, knowing this, demanded more evidence in support of the so-called Big Bang theory. Convincing evidence was found, and now the Big Bang theory is commonly accepted. There are still loopholes; a logician could no doubt find them.

There are, of course, other reasoning fallacies that are sometimes committed by the most distinguished scientists. For example, you might notice a correlation between two events A and B. They seem to occur concomitantly. How do we find out which event is the cause and which is the effect? Or does some other factor (C) cause both A and B?

A good example here is the mystery of heart disease. High cholesterol levels in the blood seem to be associated with an above-normal instance of heart disease. Many scientists think that this means that cholesterol is the cause, and heart disease the effect. Could it not be that the reverse is true? Perhaps high cholesterol is a symptom of heart disease? Could it be that both high cholesterol and heart disease are symptoms of some other, perhaps unknown factor? It is important that we consider all of the possibilities even if it seems that some of them are not likely to be borne out.

Logic is an unchallenged absolute in the world of science. We need some agreed-upon procedure for drawing our conclusions from the data available. Is it not a little disconcerting to think that we, in the enlightened twentieth century, are treating logic in the same way that the church treated their scientific theories a few centuries ago?

Some logicians have worked with so-called alternative forms of logic, where the ordinary, commonly accepted laws are supplanted with rules that seem quite bizarre at first thought. Perhaps some other logical system might work better than the one we use. Nevertheless, we have to have some foundation for scientific thought or we might end up like the old man on the mountain whom the climber met and who had come to think he sat atop the highest mountain for no apparent reason save his own narrow-mindedness and conceit. It is better for us to have a common denominator, that we can use to formulate conclusions from observed data, as opposed to simply jumping to those conclusions because they are intuitively or emotionally pleasing. Our scientific method, while at least more fair (if not better) than the ancient ways, is still not infallible. Everybody makes mistakes.

Mathematics

From logic we derive the field of pure mathematics, and all its disciplines: algebra, analysis, calculus, geometry, number theory, set theory, and topology—to name just a few. All of the theoretical or pure mathematicians adhere to the same rules of logic. Thus mathematics is, in a sense, just as absolute as the rules of logic from which it is derived, but there is a slight complication—an imperfection. Some statements are true, others are false, and a few are neither true nor false. This introduces an irresistible element of mystery into the universe. Some things simply cannot be known, one way or the other, even if we consider our logic to be perfect. Mathematicians have proven this.

The pure mathematician is an interesting kind of scientist, and one which I respect greatly. Pure mathematicians build their own universes by formulating their own axioms and employing logic to find what follows. A contradiction, if it occurs, causes the pure mathematician's universe to instantly collapse upon itself; but no matter, there are other universes awaiting exploration.

The pure mathematician does not care what others think of his or her creations, but others do think of them, and use them to achieve various ends. These are the applied mathematicians, the physicists, chemists, doctors, and astronomers. Mathematics is an indispensible part of all the sciences. Any astronomer who approaches the subject without knowledge of the mathematical basics will not get far. The power of mathematics in astronomy and cosmology was demonstrated by Isaac Newton, and later, in a most vivid way, by Albert Einstein. There are countless other examples.

Fitting Theory to Fact

It has been said that we must formulate our theories to fit the facts, and not vice versa. That appears obvious, but it is amazing how many of us, upon seeing evidence against a theory that we like, will attempt to either deny the observations or to "Band-Aid" a theory to fit the observations until that theory becomes so unwieldy as to be ridiculous.

Consider the epicycle theory of Ptolemy. Ptolemy believed that the Earth was the center of the universe, and that all of the planets, and even the Sun, revolved around our home planet. Astronomers made some observations that conflicted with this theory. Planets generally "revolved" from west to east across the background of stars. Exceptions were Venus and Mercury; they seemed to oscillate back and forth near the Sun. Even the outer planets occasionally reversed direction and mysteriously traveled from east to west for short periods. Yet, if the planets all moved around the Earth in perfectly circular paths, this retrograde motion would never be observed (Fig. 6-1). All of the planets would always move from west to east with respect to the distant stars.

Ptolemy modified his theory by postulating (and that is all it was, an assumption) that the planets followed major and minor paths around the Earth. The minor orbit, called the epicycle, was centered on, but smaller than, the major orbit or deferent. This concept is shown in Fig. 6-2. By choosing deferents and epicycles of the appropriate sizes and ratios for each planet, the retrograde motion was explained. The accuracy is quite surprising even by modern standards. The precision was improved by adding subepicycles and sub-subepicycles until one ancient king was motivated to remark that, had he been present at the creation, he might have given some advice to God!

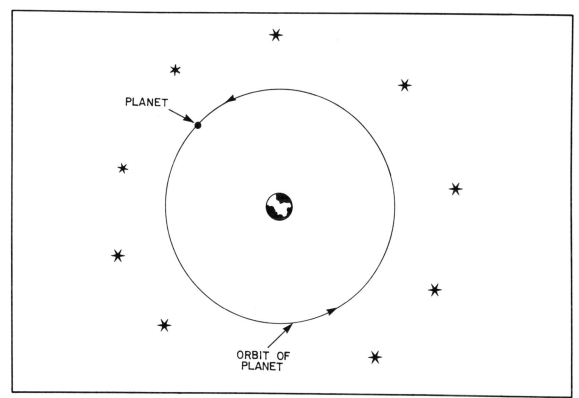

Fig. 6-1. If a planet followed a simple orbit around the Earth, the motion of the planet would always appear to be in the same direction with respect to the background of stars.

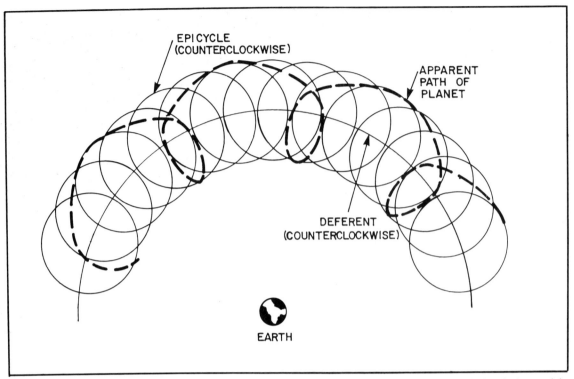

Fig. 6-2. The Ptolemaic model of the Solar System used large orbits (deferents) and small suborbits (epicycles) to explain the retrograde motion of planets.

Of course, this theory was refuted by later scientists who showed that the retrograde motion was explainable in much simpler, and more precise terms by considering the Sun as the center of the planetary system. It was some time before the heliocentric (Sun-centered) view was accepted. The scholars of the establishment did not want to give up the theory in which they had learned to believe. Their house of knowledge was warm and secure, and the new ideas dangerous because they undermined the absoluteness of Ptolemy's universe. Revolutionary ideas were thus met with hostility.

Even today we sometimes make this mistake. Somebody comes along with a theory, based on research and observation, that conflicts with established ideas. Frustrated scientists, being only human, might try to change the facts or deny that the new observations are correct. In a way, it is hard to blame them. We must have a certain system of checks and balances, lest science be reduced to a meaningless jumble of postulates. We also have to be open minded to new theories (especially those backed up by concrete observation).

Heresy

Our scientific research and theorizing is done by mortal human beings with hopes and dreams and fears that all too often interfere with rationality. Perhaps a scientist spends the greater part of his or her life working on a theory. That theory is then put to the test again and again by colleagues, and eventually the theory might fall in the face of logic. That is frustrating. Galileo put the theories of the church to the test, and those theories fell against Galileo's impeccable observations. The church got frustrated, and Galileo was made to grovel. He was branded as a heretic, a nonconformist—the kind of person likely to be stopped by the police for making an improper left-hand turn.

Having eleven times the diameter of the Earth, Jupiter is the largest planet in the Solar System. A strange spot, noticeable even with a moderate-sized telescope, marks the surface. The spot appears to be a massive hurricanelike storm that is as large as the Earth or the planet Venus. Immanuel Velikovsky (1895-1979) suggested that Jupiter ejected Venus in a violent eruption 35 centuries ago. Most scientists do not think Velikovsky was correct; his theory involves coincidences that defy the laws of chance. Suppose he was right, and the Great Red Spot is the scar that Jupiter still bears from that event! (NASA photograph.)

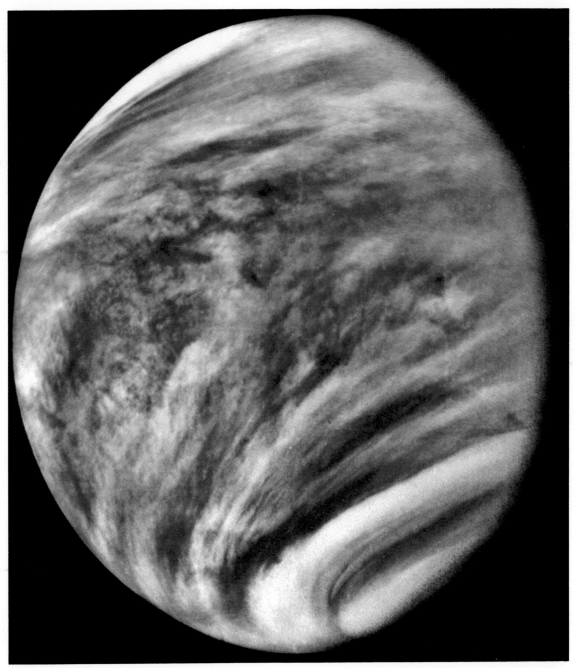

Venus, named after the ancient goddess of love—its cloud structure revealed by ultraviolet photography—is about the same diameter as our planet, but has a surface temperature of hundreds of degrees and a peculiar retrograde rotation that gives it an extremely long day. According to a theory by Velikovsky, this planet passed so close to the Earth at one time that the gravitational and electromagnetic forces caused the geographic poles to shift position, great ocean waves to rush over the continents, and kingdoms to rise and fall. Could our beautiful morning/evening star-goddess have been responsible for such violence? (NASA photograph.)

At the turn of the century, physics was regarded as a science that had been almost completely mastered by humankind. There was little else, it was thought, that remained to be discovered. A man named Albert Einstein formulated a theory that shattered the accepted Newtonian principles. Almost overnight, the theories of hundreds of physicists were disproved or shown to be less than perfect. Thousands of textbooks became obsolete. People got frustrated. They did not accept the revolutionary ideas of Einstein right away; some called his theory of relativity "complete nonsense."

Should some young scientist come along and show that the theory of relativity is not quite perfect—that there is some flaw in it—what will be the reaction of the scientific community? That rebel scientist might well be labeled a heretic. Yes, that word is still used, even today, in our supposedly enlightened society. We are more medieval than we might think. We still attack people as well as issues.

The Danger of Rank

We already know what can happen to a scientist, especially an amateur scientist or a person without the standard training in a particular field, if he or she comes up with a nonconformist theory or even an idea that rubs somebody the wrong way.

I can recall a conversation in a radio station where I worked some years ago. The subject was the effect of bandwidth on the sensitivity of a radio receiver. I contended that reducing the bandwidth improves the signal-to-noise ratio as long as the information is transmitted at no more than a certain speed (say, a few words per minute). The narrower the bandwidth, I said, the less noise would get through, while the signal strength would remain unchanged until the crucial bandwidth was reached. Lest we get lost in electronic jargon, however, I won't describe the technical details any further. That isn't the point, anyway.

A visitor to the station told me, "I think you are wrong." He proceeded to explain his reasoning, but I was not convinced.

The discussion continued, in a friendly way, for several minutes. Then the visitor began to get exasperated with my stubbornness. I would not declare that I was certainly right and he was certainly wrong; what rational person can say such a thing? Finally, he pulled his ace. He had a master's degree in electrical engineering, and that was more academic merit than I had achieved. While it was not explicitly stated, the implication hung in the air like the odor of stale food. His qualifications, all by themselves, made him right and, therefore, I was mistaken. He might just as well have said that I could sustain a hundred-volt shock and not feel a thing. I would have begged to differ, and it would have been no use. He was of a higher stratum. Smarter by cosmic standards.

In general, it is the experienced scientist who makes the most important discoveries. But what is experience? And what qualifies one to call himself or herself a scientist? Is scientific merit to be determined in hierarchical form, as in the military, so that someone with a doctorate can actually tell someone seeking a bachelor's degree how to think?

Let us hope not! Let the shoulders of even the most qualified, educated scientists periodically be shook so that their minds will not get paralyzed by conceit.

A MODERN "HERETIC"

It is a fascinating, almost unimaginable tale of cosmic disaster that changed the whole course of history on this planet. A massive object of some kind was ejected from the planet Jupiter, careened toward the Sun, passed by the Earth, and caused great astronomical and geological events. Kingdoms fell; the climate was altered; great waves flooded the continents; the Sun stood still in the sky; and fiery hail stones rained down from the heavens. The object, as it passed near the Earth, appeared as a comet with the brilliance of the Sun and the object's tail arched across more than half of the sky in full daylight.

The comet passed near the Earth a second time, a half century later, and caused further catastrophes and terror.

The interaction among the gravitation and electrical charge in the comet—and in Mars and Earth—resulted in changes in the orbit of our planet. The poles changed position. Mars, deflected from its orbit, came close to the Earth and produced tidal and electromagnetic upheavals. This all happened within the time of recorded history—about thirty-five centuries ago. The "comet" ultimately became the planet Venus (Figs. 6-3A and 6-3B).

Are you able to believe this? This story is told in ancient scriptures in many cultures. A man named Immanuel Velikovsky has said that the wild tales are more than legend; they are fact, and he provides evidence to prove his point. About 3,500 years ago, astronomical events changed history. The Solar System was cast into a state of disorder, and the Earth fell victim.

Velikovsky's theory is unfamiliar to younger students. His ideas are not usually presented in astronomy courses at the high-school or undergraduate college level. In 1950, Velikovsky's theory took the form of a best-selling book called *Worlds in Collision*.* The reaction of the public was amazing. His book became a best-seller almost overnight. The scientific community was less enthusiastic. What followed is difficult to believe for those who want to think we live in an enlightened age. Today's students are not informed of Velikovsky's theory because it has been all but censured, and Velikovsky, who died in 1979, was labeled as a "heretic."

The evidence for and against Velikovsky's theory has been documented.**Therefore, arguments will not be presented here. While I find much of Velikovsky's theory difficult to accept—it seems that the laws of chance are stacked against many of his claims—some of his general ideas have gained credence within the short time since *Worlds in Collision* was published. Perhaps the most significant is the theory of catastrophism. The universe is beseiged by violent events that have a profound, if not actually total, effect on its fate. The "heretic" Velikovsky has made an undeniable contribution to scientific thought. It is possible, even likely, that the full magnitude of that contribution will not be realized for generations. Whether or not Velikovsky was right in every detail (and almost certainly, like any scientist, he was not), he shook the shoulders of our minds. He deserves credit for that.

CATASTROPHISM

The best example of catastrophe in our universe is the Big Bang. Presumably, several billion years ago, an unimaginably violent explosion took place, distorting time and creating matter and energy. Some scientists think the Big Bang came from the remnants of a previous universe; others think it was formed from nothing. Theologians might say it came from God. We do not know who is right; perhaps no one. (The theory of the Big Bang is described in some detail in Chapter 1.)

Things tend to occur in bunches, in sudden spurts, and not in regular order. This can be demonstrated quite nicely by flipping a coin a few dozen (better yet, a few hundred) times. You will notice that the coin tends to land "heads" and "tails" in series of three or four at a time quite often. It won't take you long to get a string of five or six in a row. If you work at it for a while, you'll get a string of eight or 10. I don't spend a large part of my time flipping coins, and I have had a set of 14 "tails" in a row.

The probability, in theory, of a coin coming up "heads" or "tails" is 50-50 on any given toss. The change of getting two identical faces, say "heads," on two consecutive tosses is one in four. The chance of getting three in a row is one in eight. The probability decreases by powers of two: 1/2, 1/4, 1/8, 1/16, 1/32, 1/64, and so on. For my 14-in-a-row experience, I was working against odds of 1/16,384. I don't think I have spent that much time flipping pennies. Did I just get lucky? Was it a coincidence?

We have all had experiences wherein one remarkable event follows on the heels of another.

---

*Velikovsky, *Worlds in Collision*, Doubleday & Co., 1956. (Original edition published by Mac-Millan, 1950).
**A good summary can be found in *Scientists Confront Velikovsky*, Cornell University Press, 1977.

Fig. 6-3A. Venus as it would appear without cloud cover. Topography has been exaggerated. This view is of half of the planet from a few thousand miles away (courtesy NASA).

Fig. 6-3B. A closeup of part of the surface of Venus without cloud cover. The solid nature of the surface, and the evidence of impact craters, suggests that Venus is an old world formed along with the rest of the planets in the Solar System. Scientists doubt it was ever a comet (courtesy NASA).

You win a lottery prize and then find out that you have won a brand new car in some other contest. A tornado strikes your house and, after you have rebuilt it, another tornado hits it. (I recall hearing about a man who had been hit by three tornadoes!) Unlikely events do happen. A statistician will tell you that is simply because there are so many people in the world—so many entrants in a lottery, or so many severe thunderstorms going on—that some of these coincidences are inevitable. That is certainly true, but there is something else about remarkable, strange, bizarre events. They seem to have a greater tendency to occur in batches than mere statistics would dictate. Mathematically, I can't prove this proposition; I don't think anybody ever has. Nevertheless it does seem to hold. It is an observation that I have heard more than one person make.

The weather gives us some good examples of the irregularity of nature. Rain does not fall continuously, but comes in showers and storms of short duration. The average temperature of a place might be 50 degrees Fahrenheit, but during a heat wave it can climb to 80 or 90 degrees. During a cold snap, temperatures can fall far below zero. The average wind velocity in a town might be 10 miles an hour, but sometimes it is calm and sometimes gusts might reach 100 miles an hour.

What does this have to do with catastrophism, Velikovsky's theories, or scientific method? Bizarre coincidences are the rule in our physical universe. Velikovsky just might be correct. Scientific method is fallible even though it is the only reasoning scheme we have.

Logic is perfect by default. Mathematics is almost perfect but not quite. Physical science is far from perfect. Our universe—the Galaxy, stars, Sun, planets, and the Earth—are physical objects, the nature of which we will never totally understand. Velikovsky pointed this out relentlessly. In so doing, he upset a lot of people.

LEAPS AND BOUNDS

The tendency of events to occur in bunches, rather than in uniform fashion, can be applied to the Earth and the universe as a whole. Our planet developed in leaps and bounds, not in a regular, continuous fashion.

For example, how did life come to be on our planet? We can define life however we want. Suppose we call a molecule "alive" if it is either DNA (deoxyribonucleic acid) or RNA (ribonucleic acid). If we look far enough back in time, there was not a single DNA or RNA molecule on our planet.

We move into the future second by second, minute by minute, day by day, year by year. Eventually we reach a time where there are some DNA or RNA molecules. When did the first one form? It had to have formed sometime at some split-second instant. One second there was no life on Earth; the next second there was. The whole course of Earthly evolution was changed in a single instant of time! This eerie fact comes up however we define life.

Perhaps a more vivid way for an individual to understand this idea is found in the controversy over what, exactly, human life is. I don't want to get involved with opinions here, but there is a question. When does life begin? However we define it, we have to accept that life begins in a single instant. There is no quarter-life, half-life, or two-thirds life. You, I, and every one of us, came to life in a split second.

The course of your universe changed direction radically at a single time-point. At one second before eight o'clock in the morning on July 5, 1983, John Doe did not exist. At eight o'clock he did. Perhaps John Doe will someday be president of the United States.

Mathematicians will recognize this logic as an expression of the theorems of least upper bound and greatest lower bound. If a statement is false and then becomes true, it must change state at a single instant. There is no in-between condition. And theoretically, it should be possible to find that instant.

If we define life as anything having DNA or RNA—substances which themselves are rigorously definable—then life came on to the Earth at a point of time that we can, at least in theory, locate with great precision. The same is true of the creation of any animal or human being. All we have to

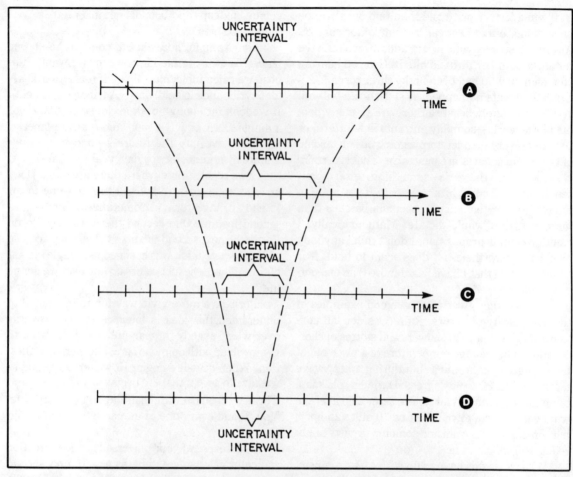

Fig. 6-4. When identifying the precise amount at which an event occurs, there is always an uncertainty interval. The size of this interval depends on the accuracy of our measurements and observations. In this illustration, the accuracy improves from A to B, B to C, and C to D.

do is close in on the time point from before and after (as shown in Fig. 6-4). The accuracy of our determination is limited only by our ability to observe and interpret.

However you define the creation of a human life—the moment of conception, the time of birth, or some time in between—you can find out from your medical records when you come to be (with an accuracy of plus or minus a few days). With the help of your parents, you might narrow it down to a few hours. Perhaps someday we will know when life began on Earth, to within one year, one day, or one second!

Other events happen in the same way: instantly. A huge meteorite smashes into the Earth. The course of history is altered within a split second. The apparition of a comet convinces some ruler to declare war on a neighboring empire. He makes his decision final with one sweep of his scepter (and perhaps a hefty swig of distilled spirits). As a result, a kingdom falls, languages and cultures change, and the course of human history is altered.

Millions of miles from the Earth, something happens in the mysterious interior of the planet Jupiter, setting up the sequence of events that ultimately leads to the violent expulsion of a massive comet. At one moment, there is no precondition for it. A millionth of a second later, a molecule has moved into the correct position to start the chain reaction culminating with the formation of the Venus-comet, and all of the resulting catastrophes that befall the Earth.

This last example is, of course, controversial because it uses the Velikovsky's theory as its basis. No one has yet proven, with mathematical perfection, that the event did not take place. We do know that catastrophic events must inevitably happen. That can be proven by pure scientific reasoning. And the strange tendency of coincident events to occur close together—the proposition I confess I cannot prove—is exactly what Immanuel Velikovsky is asking us to accept in his theory.

It is enough to make me want to believe Velikovsky. I want to think that there will always be aspects of the universe that we cannot explain. I don't want to know everything, for fear I will end up like the old man, the climber, and the young boy whose minds rotted on mountain peaks.

ORDER AND DISORDER

Albert Einstein believed that the cosmos is orderly, governed by universal laws that we, given enough time and patience, should be able to find. Although Einstein did not prove it, he believed in the so-called unified-field theory. This theory states that there exists a connection among all of the different kinds of forces in the universe.

Of course, certain factors do govern the behavior of things. If we heat a piece of wood to a certain temperature, it catches on fire. If we place a large enough potential difference between two objects, a spark will jump between them. At least, this is what has happened, as far as we know, every time somebody has heated wood or placed opposite electric charges on objects near each other. But there are things in the universe—fundamental things—that we cannot determine with certainty. Where will the electron in a hydrogen atom be at a specified instant? Its position cannot be pinpointed; all we can say is what the probability is of its being within a certain volume of space. This is shown in Fig. 6-5.

In the early and mid-twentieth century, physicists unraveled a new theory concerning energy and matter. Einstein's famous equation $E = mc^2$ provides the mathematical relation between energy and matter. Other scientists found that matter is made up of countless tiny particles, and that light, and all electromagnetic energy, is radiated in discrete packages called photons. The cosmos is a vast potpourri of particles, all flying around in unimaginably complex ways.

Billions of photons enter your eyes each second as you read this. Billions of tiny catastrophes occur as the photons land on your retina. If you could slow time down and somehow watch the photons raining down, you could count their frequency and say that, within a certain span of time, you might expect a certain number of photons to strike (Fig. 6-6). You could never predict precisely when the next one would arrive. The photons fall at random.

On the scale of the atom, the electron, proton, neutron, quark, photon, and all of the other particles that comprise our universe, things are governed by chance. This is the essence of the quantum theory. Einstein's orderly universe is reduced to chaotic randomness. If the smallest particles—of which everything is made—are ruled by nothing but probability, then everything in the universe is a product of blind chance, too. If we heat up enough pieces of wood, eventually one of them will refuse to burn. If we charge enough pairs of objects, sooner or later a spark will fail to jump the gap between them. It may take trillions of pieces of wood, trillions of charged objects, and many millions of years, but eventually the seemingly impossible must happen.

It is not possible, then, no matter how improbable it might seem, that Jupiter could have disgorged a comet that ultimately became the planet Venus?

SCIENCE AND REASON

Just because something seems to be true, is intuitively appealing, or sounds fascinating does not prove that it is reality. The tendency of coin-

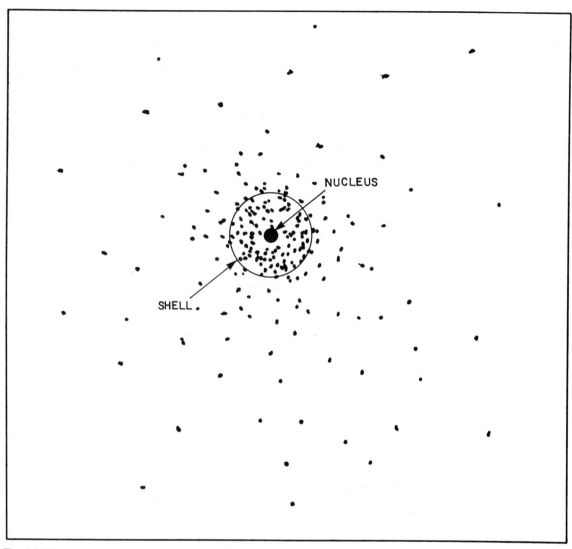

Fig. 6-5. We cannot say exactly where a particular electron in an atom will be at a given instant. We can only express the probability that it will be in a given volume of space. In this dimensionally reduced illustration of a hydrogen atom, the small dots represent the electron at various instants of time. Half of the time, the electron is within the spherical shell; half of the time it is outside.

cidences to take place more often than they should might be nothing but a illusion. We might never know. Logicians have shown us that there are things, at least in mathematics, that can be neither proven nor disproven. Quantum theorists have shown us that there are things in the universe that cannot be precisely determined. Velikovsky tells us that a comet came out of Jupiter and changed the course of Earth history.

Velikovsky's theory was met with great zeal by some and with derision or anger by others. For some, his ideas provided an astronomical explanation for biblical events. Depending on the religious inclination of a particular person, this might either please (the agnostic) or upset (the biblical literalist). For some people, Velikovsky was a heretic, mixing mysticism with pure science and tarnishing the credibility of all astronomers. The resulting upheaval in the scientific and academic world was at least as catastrophic, proportionately, as the events which Velikovsky claimed had occurred in the Solar System.

History is full of examples of professors ousted from their positions because they did not follow the orthodoxy. For many it is "conform or perish." For the outsider, whose specialty is not in the field, the pressure takes the form of derision and possibly even legal action. Giordano Bruno was burned at the stake for his ideas. Galileo was put under house arrest.

Even pure mathematicians are not immune from the inquisition of science orthodoxy. Georg Cantor, the famous set theoretician who discovered the existence of transfinite cardinals—different kinds of infinity—was mistreated by the establishment and died insane. Many of his contemporaries must have thought Cantor insane. How could there be different degrees of the infinite? When I first heard about the "alephs," and that there could be things larger than the infinity of the counting numbers, I could not believe it. It actually made me frustrated to think that there could be anything larger than infinity! The academic world of Cantor's time must have felt the same way. Now his

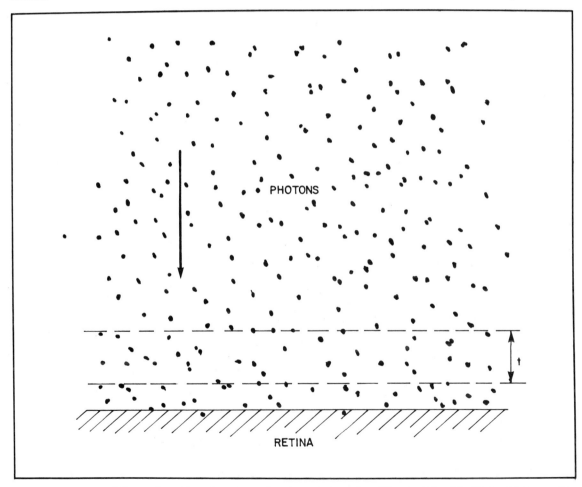

Fig. 6-6. As photons fall onto the retina of your eye, they impact at random. We can estimate with good accuracy the number of photons that can be expected to fall within a certain interval of time t, provided that interval is long enough.

theories are accepted.

New ideas, especially if they go contrary to the prevailing notion of intuition and common sense, cause frustration because they undermine the foundation of our thoughts, leaving our minds, as it were, unsheltered in the cold rain.

DEMOCRACY IN SCIENCE

Differences of opinion are necessary if progress is to be made. A fundamental example of this can be found in the government of the United States. Republicans and Democrats differ sharply on many issues. Within either political party there is some disagreement. As we all have seen, controversies often grow heated, but everyone listens to everyone else. We take all ideas into account. This is how our system arrives at solutions to problems. It doesn't work perfectly, but it functions better than other methods we have found.

Suppose there was no difference of opinion in government, and that everyone agreed on every issue all of the time. Where would that lead us? Rapidly down some road to some end, where efficiency would be almost 100 percent. The result would be tyranny and a system closed to new ideas. Dissidents would be suppressed if not prosecuted outright. We have seen many examples of such totalitarian regimes throughout history.

The same holds true in the scientific world. There must be various points of view, even widely differing ones, and they must all be taken seriously. Velikovsky, in the field of astronomy and cosmology, might be compared to an "independent" in the governmental system of the United States. He was allowed to speak his mind, because, in this country, everyone is. His ideas were rejected by many, if not most, of the people involved in science and astronomy, and that is their right. His theory was welcomed by others, heralded as a breakthrough in the quest for a link between pure science and religion. The believers, too, have their right to think as they please and speak their minds.

Some people tried to censure Velikovsky, but they did not succeed. The general public received his work with great enthusiasm. Why? Could it be because many people like change, new and different ideas, and unorthodox thought? There are strange things going on in this universe: extrasensory perception, out-of-body travel, strange creatures (such as the Loch Ness Monster) living in lakes and oceans. Are all of these phenomena just myths and legends? Some scientists want to think so; other welcome the strange and inexplicable as a challenge to further human knowledge.

Differences of opinion breed progress. Complete agreement brings stagnation. Velikovsky said that should his theories ever gain general acceptance, he would not want them to become dogma.

FOR THE SAKE OF ARGUMENT

Just for the sake of argument, let me put forth a few interesting ideas for you to think about. I'm not claiming either to believe any of these propositions or not to believe them. I don't have enough scientific evidence to make a personal decision one way or the other, but certain people have made claims for each one of these propositions—not just blind postulating, but tentative conclusions based on concrete observation.

None of these propositions has been proven irrefutably by scientific method, but the legends persist. Suppose that someday we are able to show that one or more of these bizarre tales is true!

The Bermuda Triangle

An approximately triangular geographic area in the Caribbean Sea has become famous for strange incidents. Ships and airplanes have been lost, apparently for no reason, with alarming frequency. This region is responsible for more unexplainable incidents than any other equal-sized area in the world.

Various theories have been put forth in an attempt to explain the phenomenon. Perhaps there is a powerful magnetic field in the area, resulting from terrestrial magnetism or perhaps the previous fall of a huge iron meteorite. Some people have suggested that there is a time warp in the Bermuda Triangle, and that people are pulled into other dimensions. This idea is based on the fact that some boats and planes have disappeared without a trace.

I hear laughter!

But I will think twice before I sail a small boat or fly a small plane into the Bermuda Triangle. I'll get just a little bit nervous even if I am aboard a large ocean liner or commercial plane passing through the region.

The Lost Continent of Atlantis

There is a legend about a continent that once existed in the Atlantic Ocean, somewhere in what we now call the Northern Hemisphere. The exact location is not certain. Greek mythology places a large island in the western sea, and Plato describes Atlantis as a utopia destroyed by an earthquake. One recent theory holds that Atlantis was part of the Aegean island of Thera. According to the legend, the whole continent disappeared in a catastrophe under the ocean several thousand years ago, and all of the inhabitants who remained were killed. What could have caused an entire continent, supposedly about the size of Greenland, to sink under the sea?

Maps have been made of the ocean floor, and there is no immediately obvious evidence of a continent ever having existed in the middle of the Atlantic. But geological changes might have wiped out most, or all, of the traces, leaving us with no direct physical evidence.

If there ever was a continent Atlantis, various catastrophic events could have occurred to destroy it. An asteroid might have hit the Earth; there is a deep trench in the Caribbean that some scientists think could have been dug by such an impact. The resulting volcanic and seismic disturbance might have been enough to alter the face of the Earth significantly. A small, island continent might fall, and new mountains rise in other places. The mid-Atlantic ridge, the longest mountain range in the world, if we count underwater mountains (Fig. 6-7), gives vivid proof that the Earth's crust has been, and perhaps still is, dramatically changing.

Another possible cause for the disappearance of Atlantis is—dare I say it?—that a massive celestial object passed close to the Earth one or more times, producing great tidal forces. A possible shift of the poles might have taken place; the oceans would most certainly have flooded low-lying land masses.

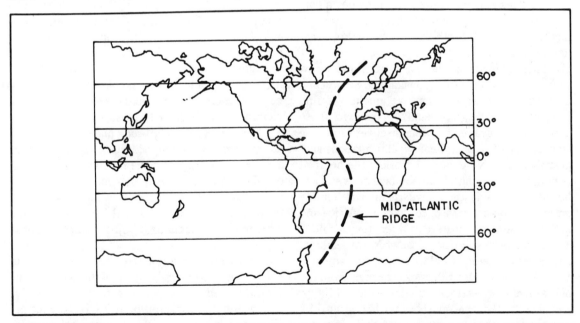

Fig. 6-7. The mid-Atlantic ridge, shown by the dotted line in this Mercator map of the world, is the longest mountain range on our planet.

It is also possible that Atlantis disappeared because of events originating solely inside our planet. We know very little about the interior of the Earth, but geologists are fairly certain that our planet is not solid all the way through. Much, perhaps most, of the volume of our planet is liquid. Volcanoes sometimes remind us of this at terrible cost of life and property.

The most likely conclusion that science will reach, it seems to me, is that there never was an Atlantis, or that its existence can be explained in terms of present geography (maybe the British Isles or Iceland or possibly even Greenland were that continent). We must remember that the past happened just one way. We can't say that "there is a 20-percent chance that Atlantis existed and then sank into the sea." Either it did or it did not. The same rule applies to Velikovsky's theory, and every event the universe has known up to now. The facts are there. It is up to us to find them.

Polar Shifts

One geographic theory, of which I am especially fond—I think it is interesting and provides a connection among numerous past events—is the idea that the geographic North Pole and South Pole were not always where they are today.

There is evidence that palm trees once grew as far north as Greenland, Alaska, and Siberia and as far south as Antarctica. We might yet find massive coal and petroleum deposits in these places. Oil has already been discovered on the North Slope of Alaska. Geological formations suggest that there were once glaciers in the heart of Africa, and that the Hawaiian Islands were ice-bound. Some scientists have theorized that the poles of our planet were once located in Africa and in the Pacific Ocean, instead of where they are now.

If that were still true, most of North America and South America would be near the equator. Europe would be about the same distance from the pole as it is now. The ancient civilizations of Egypt and Mesopotamia would be farther from the equator than they are today, and their climates would probably be cooler and wetter. Antarctica would be a tropical, steaming jungle continent. The Arctic Ocean would have been a subtropical sea.

Imagine living in Minneapolis and watching the sun rise in what we now call the north, course directly overhead, the palm trees casting their shadows on the wet, hot ground, the dinosaurs roaming among the lakes and streams. Imagine the brilliant, fiery red sunset in what we now call the south, and the constellations traveling an unfamiliar course. Think of the Zodiac slanting sideways in the heavens compared to what we think of today. Think of a map of the world with the North Pole in the middle of the Pacific and the South Pole in Africa!

Much of the equator would be marked by land mass rather than by water (as is the case now). One polar region would lie in a vast, broad ocean, much larger than the present Arctic Ocean. The oceanic currents and tides would certainly be different than they are now. The atmospheric circulation also might be different. Places now covered by desert might have been fertile farmland, and many of today's lush jungles would have been snowy tundra.

The polar shift might have taken place slowly, over millions and millions of years, while the continents gradually changed their positions as the Earth's crust floated over the mantle. There is indeed good evidence—the continental drift theory—to suggest that continents do move. In a few hundred million years, the geographic North Pole might be somewhere in Europe and the South Pole in the vicinity of the Society Islands.

Suppose a polar shift took place suddenly and not so long ago. How could that happen? Perhaps a massive object struck the Earth and knocked it out of its previous plane of rotation. Velikovsky suggests that it occurred because of interaction between the magnetic fields of the Earth and some other massive, magnetized object that passed near it.

As recently as a few thousand years ago, our home planet was struggling out of the grip of a winter that lasted for eons. This period, called the Ice Age, was not the only one of its kind; evidence has been gathered that there were several past Ice Ages. What would have caused them? The theories are too numerous to discuss in detail here. One possibility is the ice ages were precipitated by sudden shifts in the locations of the geographic poles, with consequent changes in ocean currents and atmospheric circulation.

I don't claim to know how or if a sudden polar shift ever occurred, but the idea ought to be considered. After gathering as much data as possible, we can then ask ourselves if it is possible that a cosmic catastrophe was responsible.

Astrology

Certainly one of the most widely believed theories—one which scientists uniformly reject—is the belief that the movements of the Sun, Moon and planets influence the course of human events. I find that my daily horoscope (yes, I read it often!) bears little, if any, relationship to what happens in my day-to-day life. There have been evenings on which I have said, "Thank heaven I didn't follow the advice of my horoscope today," and a few days on which something happened that appeared to be predicted by my horoscope.

My sentiments about astrology are summed up quite well by the words of one radio announcer I heard when I was a child: "Today is not a good day to drive a golf ball in a tile bathroom."

As a person who tries to be logical, I might suggest we consider the possibility that it is not the movement of the planets that affects us, but instead it is us who affect the movement of the planets! Remember the vagaries of cause and effect!

However implausible astrology might seem to the scientist, there are certain aspects about it that must be considered with some seriousness. First, horoscopes can become self-fulfilling prophesies. If enough people act according to their horoscopes, the chances are improved that their predictions will come to pass. Example: the ruler of Russia has a horoscope that says, "Great danger lurks far over the horizon." The president of the United States gets a similar message such as, "Storm clouds are gathering on the other side of the world." That is an extreme example, but I think it conveys the point.

The second reason we should think about astrology stems from the question: Why do so many people believe in it? One possible answer is that the belief derives from legends that have been passed down through the ages—legends that are in turn based on actual events. What events? Perhaps a comet or meteorite struck the Earth and caused a spectacular cataclysm. This would show that space can, and sometimes does, influence humankind in a direct way. Comets and meteorites do fall to the Earth; craters are undeniable evidence of that.

Before we dismiss astrology entirely, we have to make sure that we have all of the facts in hand. We obviously do not. There might have been some other, unknown type of cosmic encounter. And gravitational effects can't be ruled out.

The Chicken Eggs

The motions of the Sun and Moon definitely affect the Earth. We know this because of the activity of ocean tides. Smaller objects, too, are influenced by tidal forces. Even human cells might be affected.

Raw chicken eggs can be stood upright quite easily in late March and late September, but it is nearly impossible to accomplish this in December and June. It is unlikely that a chicken egg will ever stand on end unless it is attempted within a couple of weeks of the equinoxes on about March 21 and September 21. The equinoxes evidently have a peculiar effect on chicken eggs. Most likely it is gravitational. I didn't believe this until I tried it with a lot of eggs. Maybe you want to try it, too. Make sure the table is flat and steady, and don't put the eggs too close to the edge.

If chicken eggs, only an inch or so in diameter, are affected by tidal forces just as is the Earth, then it is not unreasonable to propose that our individual body cells—whose structures are similar to chicken eggs—behave differently near the equinoxes than near the solstices. We have all experienced mood changes in the spring and the fall. Maybe these emotional reactions are based on more than the coming and waning of the swimming or skiing seasons.

Another example of possible tidal effects on

humans is the full Moon. Some scientists have suggested that the tidal pull of the Moon, maximum at new and full phase, might affect human cells and thus human behavior. The light of the full Moon could have an additional psychological effect.

Human beings, constantly searching for correlations between human events and movements in the heavens, might be subconsciously influenced by tidal forces on their cells, and by the light and beauty of a full Moon. People might be inclined to do certain things more often than usual in late March and late September or when the Moon is full. If that is true, then movements of celestial bodies do, in fact, influence human events. Tidal forces caused by the planets, while much smaller than those of the Sun and the Moon, could also play a role. This

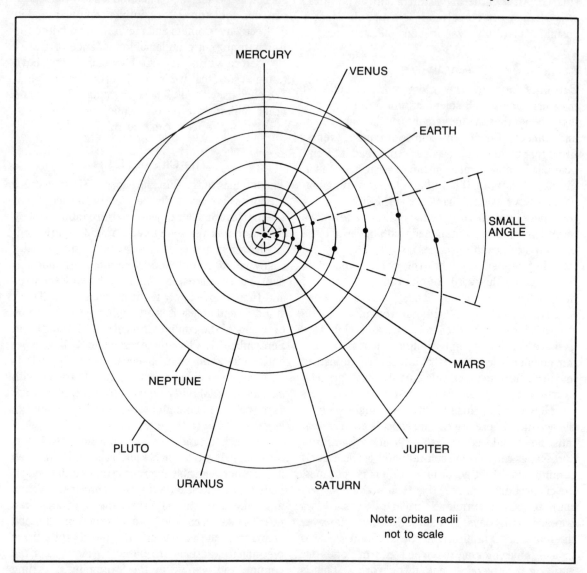

Fig. 6-8. When the planets are in close alignment on one side of the Sun, gravitational effects on the Earth are maximal. This could cause unusual human behavior, leading some ancient scientists and philosophers to conclude that planetary motion influences human events.

would be especially true at a planetary conjunction when the planets are nearly aligned on one side of the Sun (Fig. 6-8).

Astrology might, after all, share some common ground (however little) with reality.

Mental Telepathy

As a final example of a strange phenomenon to bring up for the sake of argument, consider extrasensory perception, and in particular, mental telepathy. Is there some way that brains or minds can communicate with each other without the aid of speech or electronic media? Many people have had experiences that make them absolutely convinced that this has happened.

Laboratory experiments have been conducted in attempts to prove or disprove the hypothesis that telepathy occurs. Various methods have been tried, and the apparatus set up to avoid any possibility of subconscious cues having an effect. Numbers, cards, letters of the alphabet, and other symbols have been used as the information to be "sent" and "received." Most of these experiments have turned out negative. Messages are not "received" correctly any more often than sheer chance would dictate.

But there is a problem with these conclusions, and it represents one of the worst (and one of the easiest) mistakes that can be made in the implementation of scientific method. That is the drawing of a conclusion based on insufficient data.

We never really know, for sure, that we are examining all of the factors that exist. The universe is too complicated. We can only be reasonably certain, quite convinced, or almost positive about something. To make absolute claims is a dangerous business.

Consider the following scenario. A man gets a terrible feeling that something has happened to his brother, with whom he is very close, although they live a thousand miles from each other. The man calls home on an impulse and hears the news; his brother has been injured in a car accident. His mother is crying on the phone and his father is practically in shock. The emotion hangs in the air like static electricity. The man is convinced that he experienced mental telepathy from his brother, mother, and father.

This sort of thing happens all the time.

The laboratory experiments for telepathy are conducted in sterile, controlled environs, but the real universe is not sterile. We cannot ignore emotion when we conduct experiments for telepathy. It might well be that emotion is the main factor or only one of the factors involved. I suggest that some future telepathy experiments be based primarily on emotion rather than on dry numbers, letters and symbols. The results might be surprising.

Any scientific theory is open to question. Observers improve their techniques, always bringing in new data for consideration. One observation can alter an entire theory dramatically. One experimentalist can keep a dozen theorists busy.

All theories deserve attention. The facts are absolute—what is true is true—but our knowledge of truth is a likeness that can approach, but never reach, perfection. It is possible that there are some things about this universe that we will never know no matter how sophisticated our observing equipment becomes and regardless of how powerful our theorizing methods get. Anything is possible; every conclusion is tentative.

DO COMETS BRING LIFE?

One of the more extreme aspects of Velikovsky's theory is that the terrible Venus-comet rained carbohydrate substances ("manna") onto the Earth as it passed by. According to spectral observation, carbon and hydrogen do exist in comets. Hydrocarbons have been found in the atmospheres of both Venus and Jupiter as well. This is part of the reason Velikovsky thought that Jupiter ejected the Venus-comet. Both planets, and comets in general, all contain similar substances. Perhaps he forgot that such materials were common to all the ancient stars that exploded, and thus we should naturally expect that we would find such matter in all kinds of celestial objects.

Hydrocarbons (of which carbohydrates can be considered a special form) are one of the essential ingredients for life. All of our cells, and the cells of plants and animals, contain these substances. Hydrocarbons can take many forms, including simple sugars like glucose, fructose, and sucrose. Starches and cellulose are groups of simple sugars attached together. Fats include such things as vegetable oil, butterfat, and lard. These molecules all consist of carbon, hydrogen, and oxygen atoms combined in various ways. There are also other types of hydrocarbons such as alcohol, gasoline, and refined oil. Our bodies can derive nutrition from some of these, but not from others. You will get usable calories from starch, but you cannot digest cellulose. Gasoline will make you sick.

If a comet passes very near the Earth, or impacts on it, we might logically expect that the carbon and hydrogen, oxidized by our atmosphere, would form various sorts of carbohydratelike substances. Velikovsky theorized that, as the Venus-comet passed near the Earth in ancient times, edible carbohydrates rained down onto the Earth, sustaining people through the terrible times that ensued. I doubt that people could subsist for very long on pure carbohydrate. Protein, vitamins, fatty acids, and minerals are also needed for the sustenance of human life. I am skeptical of Velikovsky's specifics on that score, but he brings up an important issue that ought to be considered.

Comets are believed to contain ammonia, and ammonia is a compound that has nitrogen as well as carbon, hydrogen, and oxygen. Nitrogen is necessary if proteins are to form, and proteins are in turn essential for life to exist as we know it. Nitrogen is also abundant in the Earth's atmosphere. While we cannot expect that complex proteins exist in comet nuclei, and we cannot jump to the conclusion that proteins would immediately form from hydrocarbons in the Earth's air, it is possible that simple molecules—the predecessors of proteins—formed during a comet or meteorite impact.

Comets and meteorites could be directly responsible for the development of life on this planet!

Biologists think that life on our planet probably resulted from the formation of amino acids as a result of heat. Certain inorganic compounds, heated to a sufficient temperature, become the essential ingredients of life. If this is how life originated on Earth, what caused the heat? Lightning has been suggested as one possibility. Thundershowers, it would seem, were common in prehistoric times and even in the eons before life existed on this planet. A flash of lightning can briefly raise the temperature to several thousand degrees Fahrenheit in a small area.

Perhaps something else caused the heat necessary for the formation of amino acids. It might have been the impact of a large comet or meteorite. A massive object from space would heat up the Earth over a vast region—perhaps many thousands of square miles in area—while a lightning stroke produces high temperatures within a surface area of just a few square inches.

If the chemicals needed for the formation of amino acids were not already present on Earth, a comet could have brought them. The heat of friction as the nucleus fell through the atmosphere would be sufficient to change some of the comet's icy compounds into amino acids. Therefore, a comet could have not only supplied the heat, but also the chemicals themselves needed to start the chain of life.

Suppose a large celestial fragment crashed into the Earth and heated up a thousand square miles of ocean surface enough to allow the formation of amino acids. That would have to be a large bolide, but space is littered with comets and meteorites big enough to produce that kind of event. The Tunguska Event, with all its consequent destruction, was evidently caused by only a tiny piece of a comet.

A lightning stroke generally has a surface effect covering about 20 square inches. We are talking about an area ratio of 200 billion (200,000,000,000) to 1! In such a case, it would take 200 billion lightning strokes—that's a lot of thundershowers—to have the same surface effect as a single large comet or meteorite impact. If we consider that amino acids could have formed in the air, we must deal with a ratio of volumes that is even larger than the ratio of surface areas!

Of course, we cannot say how many comets and meteorites have hit the Earth (especially during prehistoric times), but they do hit. Given enough millions of years, it is statistically almost inevitable. It would only take one.

It might be that the first rudimentary forms of life developed in some ancient ocean as a result of a cosmic collision. Amino acids rained down from the sky and fell on water and on land, or were formed from Earth-based compounds in the intense heat of impact. DNA, RNA, cells, and tiny plants followed. Then came the small marine animals, the fish, and the land animals. Charles Darwin has theorized the rest of the process.

Imagine that you and I might owe our existence to a single meteorite or comet! We might never be able to prove it but we'll probably never disprove it, either.

Epilogue

SHORTLY AFTER CHRISTMAS, 1984, SCIENTISTS put an artificial comet into space. It was made of barium and was launched from a West German satellite 70 thousand miles above the surface of the Earth. The experiment was made in order to study the effects of the solar wind and its interaction with the magnetic field of planet Earth.

As the material ejected from the satellite, the barium began to evaporate into space. The comet nucleus turned greenish, and then purple as the tail began to form—blowing along with the solar wind.

At first, the object was too dim to be seen from the ground without the aid of binoculars, but finally the ethereal man-made comet became visible to the naked eye. Its brilliance and size outstripped expectations. From tracking aircraft, the tail achieved a span of about 10 degrees of arc in the sky. There were two aircraft that carried out most of the observations; one plane took off from San Francisco and the other from Tahiti.

Dr. Gerhard Haerendel, coordinator of the experiment and director of the Max Planck Institute for Extraterrestrial Physics in West Germany, explained that it would take some time to evaluate all of the data. He expressed hope that more experiments could be conducted in the future, under different solar-wind conditions and different Earth-Sun orientations. More launches have been scheduled.

Larger artificial comets will almost certainly be launched in the future, giving scientists and astronomers greater insight into some of the most puzzling mysteries of the Sun and the Solar System. Someday, astronauts might have the opportunity to manufacture comets similar in structure to real ones. The comets could be placed in permanent solar orbit to give us a constant gauge of the intensity and direction of the solar wind.

For scientists, permanent man-made comets will be an indispensable tool. Most of the rest of use will only notice that comets are more frequent, and less mysterious, wayfarers of space than in the past. Perhaps to our children and grandchildren, comets will be an everyday sight.

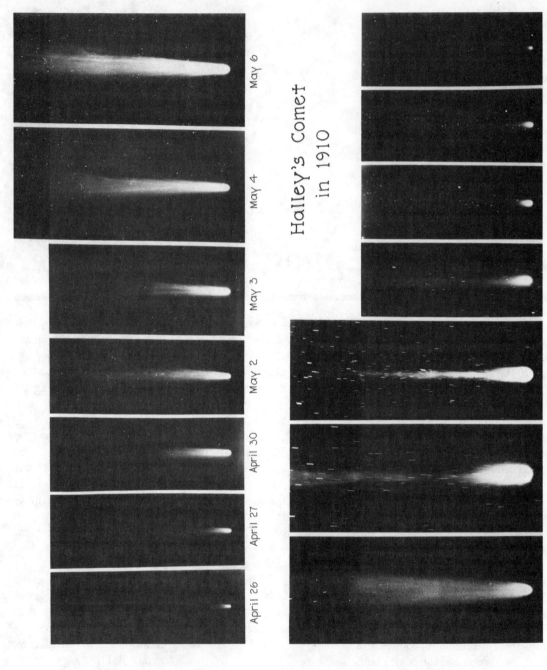

Halley's Comet as seen on 14 occasions from April 26 to June 11, 1910. The tail was most spectacular during May (courtesy of Mount Wilson and Las Campanas Observatories, Carnegie Institute of Washington).

Bibliography

Ash, Russel and Grant, Ian *Comets: Earth's Most Mysterious Visitors from Space*. Bounty Books, 1973.

Brown, Peter Lancaster *Comets, Meteorites & Men*. Taplinger Publishing Company, 1974.

Gibilisco, Stan *Black Holes, Quasars and Other Mysteries of the Universe*. TAB BOOKS Inc., 1984.

Goldsmith, Donald, Ed. *Scientists Confront Velikovsky*. Cornell University Press, 1977.

Hunt, Garry E. and Moore, Patrick. *The Planet Venus*. Faber and Faber, Limited, 1982.

Le Maire, T. R. *Stones from the Stars*. Prentice-Hall, Inc., 1980.

Maffei, Paulo *Monsters in the Sky*. Avon Books, Division of Hearst Corporation and Massachusetts Institute of Technology Press, 1980.

Mason, John, and Moore, Patrick *The Return of Halley's Comet*. W. W. Norton & Company, 1984.

Sagan, Carl *Cosmos*. Random House, 1980.

Seargent, David A. *Comets: Vagabonds from Space*. Doubleday & Company, Inc., 1982.

Velikovsky, Immanuel *Worlds in Collision*. Doubleday & Company, Inc., 1956.

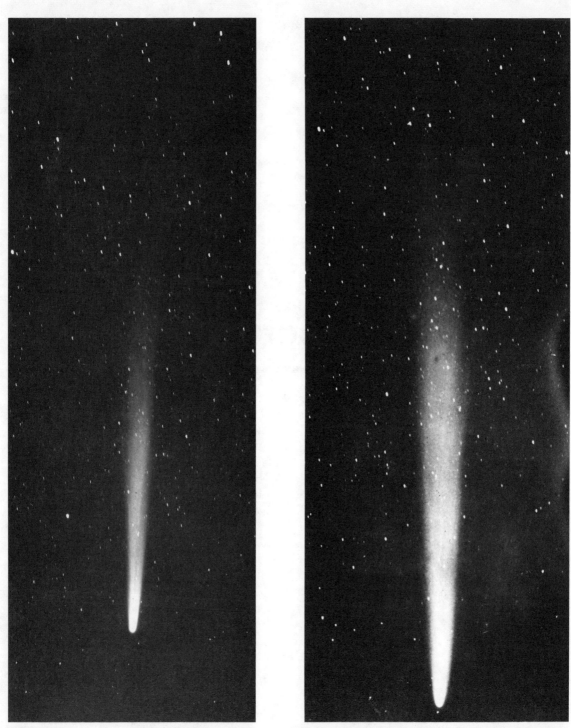
These views of Halley's Comet were taken in 1910 from Honolulu, Hawaii with a 10-inch-focus Tessar lens. At left, on May 12 the tail is 30 degrees long. At right, on May 15 the tail is 40 degrees long.

Index

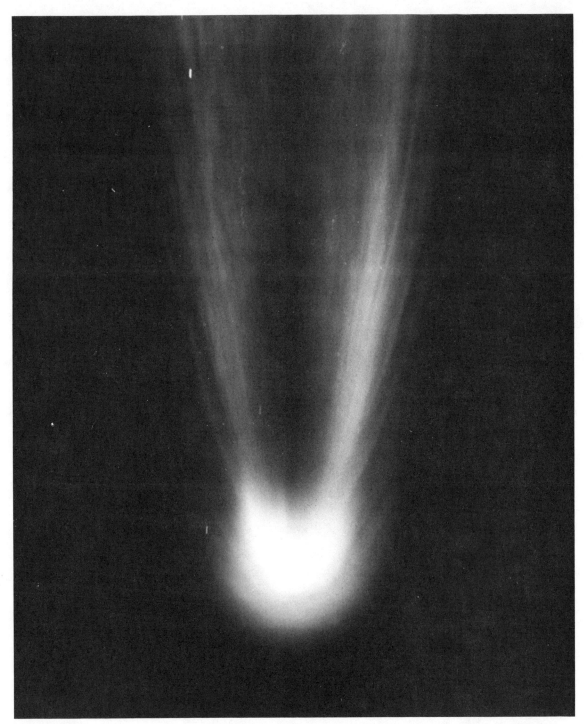

The head of Halley's Comet as seen on May 8, 1910 (courtesy of Mount Wilson and Las Campanas Observatories, Carnegie Institute of Washington).

Index

A

Adonis, 15, 16, 168, 169
Aldrin, Edwin, 144
Andromeda galaxy, 120, 123
Andromedids, and Biela's Comet, 157
Angstrom units, 46
Apollo, 15, 16, 168, 169
Arend-Roland, 81, 83, 85, 86
Aristotle, 28, 42
Asteroid Belt, 15, 55, 64, 162, 166, 168, 171
asteroids, finding and tracking, 169
asteroids, 14
astrology, 193
astrophotographic comet searching, 127
astrophotography, 125
Atlantis, lost continent of, 191

B

Barringer crater, 150
Bennett, Comet, 82, 97
Bermuda Triangle, 190
Betelgeuse, 110
Bible, the, 2
Biela, Comet, 80, 157
Biela's Comet and Andromedids, 157

Big Bang, the, 1-5
binoculars, 111-12
Bode numbers, 167
Bode, Johann, 14
Brahe, Tycho, 78
Bruno, Giordano, 189

C

Caesar, Julius, 28
Callisto, 143
Canis Major, 58
Cantor, Georg, 189
Capricorn, 101
Cassegrainian reflector telescope, 113
Cassiopeia, 119, 122
catastrophism, science and heresy, 175
Ceres, 15, 168, 173
Comet of 1882, Great, 81
Comet 1843 I, 78
Comet 1862 III, 81
Comet 1910 I, 81
Comet 1914 V, 81
Comet 1947 XII, 81
comet apparitions, 63
comet apparitions, early, 73

Comet Arend-Roland, 76, 81-83, 86
Comet Bennett, 82, 97 Comet Biela, 78, 80, 157
comet cloud, the, 25
Comet DeChesaux, 76
Comet Encke, 157
Comet Ikeya-Seki, 65, 75-76, 82, 97
Comet IRAS-Araki-Alcock, x
Comet Kirch, 1680, 78
Comet Kohoutek, 50, 55-57, 87, 92, 94, 95, 96, 131
Comet Mitchell-Jones-Gerber, 121
Comet Mrkos, 82, 88, 90, 91
comet nuclei, 49
comet nuclei, rotation of, 66
Comet of 146 B.C., 77
Comet of 1744, Great, 78, 79
Comet of 1861, Great, 76, 81
Comet of 1882, Great, 81
Comet of 240 B.C., 77
Comet of A.D. 1066, 77
Comet of A.D. 1077, 77
Comet of A.D. 1106, 77
Comet of A.D. 1222, 77
Comet of A.D. 1264, Great, 77
Comet of A.D. 1402, 78
Comet of A.D. 1472, 78
Comet Skjellerup, 1927, 81
comet spectroscopy, 48
comet swarm, 27
comet tail, 52, 90
comet tails and the solar wind, 51
comet tracking, 131
Comet West, 43, 92, 97, 131
Comet, Julius Caesar's, 77
comet's head, visit to a, 104
comets and life, 195
comets and meteor showers, 157
Comets, "great" 58
comets, basic theories for origin of, ix
comets, brightness of, 56
comets, characteristics of memorable, 74
comets, dirty-snowball model of, 66
comets, early theories about, 42
comets, hunting for, 109
comets, ionized tails of, 160
comets, longevity of, 70
comets, Lyttleton meteoroid-swarm model, 40-41
comets, naming of, 76
comets, new or old, 124
comets, orbits of, 55, 58, 63, 68, 99
comets, searching for, 107, 117

comets, shape and appearance of, 52
comets, spaces probes to, 103
comets, sungrazer, 54
comets, the fire theory for origin of, 36
comets, the ice theory for origin of, 36
comets, the most famous, 76
comets, the parts of a, 39
comets, Whipple dirty-snowball, 40
comets, why search for, 131
crater erosion on Mars, 154
crater Tycho, 167
crater, Gulf of Mexico, 151
crater, Hudson Bay basin, 150
crater, James Bay, 151
craters, 136, 138
craters, homemade, 142

D

Daedalus, 15
DeChesaux's Comet, 76, 78
Deimos, 18-20

E

Earth, craters on the, 136, 145
Earth, the, ix, xi, 1-2, 4, 9, 12, 15-16, 18, 18, 28, 31, 33, 34, 49, 53, 58, 64, 71, 74, 92,
Einstein, Albert, 187
Enceiadus, 136
Encke, Comet, 157
Eros, 15, 168
Eta Aquarids, 157

F

Fraunhofer lines, 47

G

galaxies, 9
galaxies, formation of, 4
Galileo Galilei, 42, 45
Gamow, George, 2
gegenschein, 164
Great Red Spot of Jupiter, 28
Greenland, 191

H

Halley, Edmond, 97, 98, 107
Halley's Comet and meteor showers, 157
Halley's Comet in 1910, 81
Halley's Comet in 1985-86, 100
Halley's Comet, xi, 37, 42, 51, 63, 70-73, 77, 119

Halley's Comet, history of, 97
Halley's Comet, orbit of, 99-100
heresy, catastrophism, and science, 175
hydrocarbons, 195
Hyperion, 21

I

Icarus, 15, 16
Ice Ages, causes of the, 192
Ikeya-Seki, comet, 65, 75-76, 82, 97

J

Jupiter and Venus, 187, 195
Jupiter, xi, 14-15, 17-18, 21, 27-28, 33, 69, 99-100, 110, 143, 153, 162, 165, 168, 171
Jupiter, Great Red Spot of, 29-30

K

Kepler, Johannes, 16, 53
Kirch, G. 78
Kitt Peak, 110
Kohoutek, Comet, 50, 55-57, 87, 92-96, 113
Kohoutek, L., 87

L

lagrangian points, 164
Lyttleton, Raymond, ix, 28, 31, 39, 49, 51

M

Markos, Comet, 89
Mars, xi, 14-15, 28, 33-34, 53, 99, 110. 136, 166
Mars, crater erosion on, 154
Mars, craters on, 136
Mars, moons of, 19
mental telepathy, 195
Mercury, 15-16, 18, 33-34, 53, 92, 136
Mercury, surface of, 137
meteor showers and fireballs, 136
meteor showers, 25, 133, 158, 156
meteoric dust, 162
meteorite impacts, Grand Canyon, 152
meteorite, photograph of a, 135
meteorites, makeup of, 161
meteoroids, makeup of, 161
meteoroids, cloud of, 67
meteoroids, orbits of, 159
meteors seen at sunset, 155
meteors seen at sunrise, 155
meteors, 21
meteors, moons, and rings, 18

Milky Way, 4-5, 7-10, 18, 36, 120, 162
Milky Way, photographing the, 128
Mimas, 136, 140
Moon, Apollo astronauts on the, 153, 162
Moon, craters on the, 136, 145
Moon, gravitational influence of the, 167
Moon, meteoric dust on the, 163
Moon, the, 18, 21-24, 34, 42, 74-75, 108, 110, 117, 173
moons, 21
moons, meteors, and rings, 18
moons, the orbits of, 22
Mrkos, Comet, 82, 90-91

N

Neptune, 17, 27, 33, 69, 98-99, 165, 171
Newton, Isaac, 97
North Pole, 192
North Pole, shift of the, 31

P

Pegasus, 101-02
perihelion, 68, 69, 131
Perseids meteor showers, 136
Phobos, 18, 19, 136, 139
photographing comets, 127
photographing stars, galaxies and comets, 126
Piazzi, Giuseppi, 14-15, 168
planets, alignment of the, 194
planets, the effects of the outer, 69
planets, theories for creation of the, 9
Pluto, 15, 18, 25-28, 34, 69, 98-99, 169, 171
Pollux, 56
Ptolemy, 56

R

rings of Saturn, 25
rings, moons, and meteors, 18

S

Sagan, Carl, 31
Sagittarius, 10
Saturn, 17, 19, 25-27, 33, 70, 99, 136, 140, 153, 171
Saturn, rings of, 26
science, catastrophism, and heresy, 175
science and reason, 187
science, democracy in, 190
scientific method, the, 177
Siberia, Tunguska, ix, 31
Sirius, 58, 110

Skjellerup, Comet, 81
solar flares, 65
Solar System, creation of the, 9, 12
Solar System, end of the, 33
Solar System, formation of the, 13
Solar System, origin of the, 134
Solar System, rotating-cloud theory for creation of the, 12
Solar System, the, 1, 4, 9, 17-18, 21, 42, 49, 64, 69, 99, 150, 165, 166, 173
solar system, tidal theory for creation of the, 9
solar wind and comet tails, the, 51
solar wind, 52, 65
South Pole, 192
South Pole, shift of the, 31
space debris, 133
space travelers, hazards for, 170
spectroscope, the, 46
spectroscopy, 48
Spica, 56
Suess, F. 166
Sun, birth of the, 9
Sun, the, 2, 4, 9, 14-18, 21, 25, 31, 33, 49, 51, 53, 55, 58, 63-64, 69, 71, 64, 68, 110, 134, 193
sungrazers, 70
supernova, 9

T

Tebbutt, John, 76
tektites, 166-67
telepathy, mental, 195
telescope, 42, 11
telescope, reflector, 116
telescope, refracting, 109, 115
telescopes, 112
telescopes, azimuth-elevation for, 118
telescopes, equatorial mounting, 118
telescopes, refractor or reflector, 113
Tietz, J.D., 14
Titan, 25, ,153
Tunguska Event, 31, 131-32, 196

U

universe, end of the, 34
universe, expansion of the, 2
universe, fundamental questions about the, 1
Uranus, 14, 17, 25-27, 33, 69, 70, 98-99, 165

V

Vega. 14
Velikovsky, Immanuel, ix, xi, 28, 187-95
Venus and Jupiter, 187-195
Venus as a comet, 196
Venus without cloud cover, 183
Venus, xi, 28, 34, 58, 75, 92, 99, 110, 136, 153, 166, 195
Venus, the surface of, 150, 184

W

weather, cloudiness and, 110
weather, humidity and, 111
weather, temperature and, 111
West, Comet, 43, 92, 97, 131
Whipple, Francis, ix, 25, 28, 31, 39, 49, 64, 66
Wilson, Mount, 110

Y

ylem, 2

Z

zodiac, 192

OTHER POPULAR TAB BOOKS OF INTEREST

The Illustrated Handbook of Aviation and Aerospace Facts (No. 2397—$29.50 paper only)

The Encyclopedia of Electronic Circuits !No. 1938—$29.95 paper; $50.00 hard)

The Personal Robot Book (No. 1896—$12.95 paper; $21.95 hard)

The Illustrated Dictionary of Electronics—3rd Edition (No. 1866—$21.95 paper; $34.95 hard)

Principles and Practice of Electrical and Electronics Troubleshooting (No. 1842—$14.95 paper; $21.95 hard)

333 Science Tricks and Experiments (No. 1825—$9.95 paper; $15.95 hard)

Discovering Science on Your ADAM™, with 25 Programs (No. 1780—$9.95 paper; $15.95 hard)

Exploring Light, Radio and Sound Energy, with Projects (No. 1758—$10.95 paper; $17.95 hard)

Second Book of Easy-to-Build Electronic Projects (No. 1679—$13.50 paper; $17.95 hard)

Satellite Communications (No. 1632—$11.95 paper; $16.95 hard)

Build Your Own Laser, Phaser, Ion Ray Gun, and Other Working Space-Age Projects (No. 1604—$15.50 paper; $24.95 hard)

$E=mc^2$: Picture Book of Relativity (No. 1580—$10.25 paper; $16.95 hard)

How to Forecast Weather (No. 1568—$11.50 paper; $16.95 hard)

Electronic Projects for Photographers (No. 1544—$15.50 paper; $21.95 hard)

Electronic Databook—3rd Edition (No. 1538—$17.50 paper; $24.95 hard)

Beginner's Guide to Electricity and Electrical Phenomena (No. 1507—$10.25 paper; $15.95 hard)

The Build-It Book of Electronic Projects (No. 1498—$10.25 paper; $18.95 hard)

How To Design and Build Your Own Custom Robot (No. 1341—$13.50 paper only)

Encyclopedia of Electronics (No. 2000—$58.00 hard only)

Time Gate: Hurtling Backward Through History (No. 1863—$16.95 paper; $24.95 hard)

Artificial Intelligence Projects for the Commodore 64™ (No. 1883—$12.95 paper; $17.95 hard)

Robotics (No. 1858—$10.95 paper; $16.95 hard)

333 More Science Tricks and Experiments (No. 1835—$10.95 paper; $15.95 hard)

Violent Weather; Hurricanes, Tornadoes and Storms (No. 1805—$13.95 paper)

Handbook of Remote Control and Automation Techniques—2nd Edition (No. 1777—$13.95 paper; $21.95 hard)

Designing and Building Electronic Gadgets, with Projects (No. 1690—$12.95 paper; $19.95 hard)

Robots and Robotology (No. 1673—$8.25 paper; $13.95 hard)

How to Make Holograms (No. 1609—$17.50 paper; $26.95 hard)

The GIANT Book of Easy-to-Build Electronic Projects (No. 1599—$13.50 paper; $21.95 hard)

Principles and Practice of Digital ICs and LEDs (No. 1577—$13.50 paper; $19.95 hard)

Understanding Electronics—2nd Edition (No. 1553—$9.95 paper; $15.95 hard)

Experiments in Four Dimensions (No. 1541—$17.50 paper; $24.95 hard)

Black Holes, Quasars and Other Mysteries of the Universe (No. 1525—$13.50 paper only)

Understanding Einstein's Theories of Relativity: Man's New Perspective on the Cosmos (No. 1505—$11.50 paper only)

Projects in Machine Intelligence for Your Home Computer (No. 1391—$10.95 paper only)

Basic Electronics Theory—with projects and experiments (No. 1338—$15.50 paper; $19.95 hard)

TAB BOOKS Inc.

Blue Ridge Summit. Pa. 17214

Send for FREE TAB Catalog describing over 750 current titles in print.